Secrets *of the* Light

Secrets of the Light

Lessons from Heaven

Dannion and
Kathryn Brinkley

HarperOne
An Imprint of HarperCollins*Publishers*

All "Reflections and Affirmations" in Chapters 15 through 21 have been excerpted from *Jewels for the Soul: Spiritual Reflections and Affirmations for the Heart and Soul* by Kathryn Marie Brinkley.

HarperCollins books may be purchased for educational, business, or sales promotional use. For information, please e-mail the Special Markets Department at SPsales@harpercollins.com.

HarperCollins Web site: http://www.harpercollins.com

HarperCollins®, ®, and HarperOne™ are trademarks of HarperCollins Publishers

FIRST HARPERCOLLINS PAPERBACK EDITION PUBLISHED IN 2009

Library of Congress Cataloging-in-Publication Data is available upon request.

ISBN 978-0-06-166246-1

21 LSC(H) 11

To the members of the Twilight Brigade:

We thank you for the tireless work and undaunted commitment you have made to being at the bedside of those leaving this world. Each of you is a divine inspiration as well as a living representation of compassion in action.

&

To Ralph Everett Coleman and Kathryn (Kat-Kat) Marie Coleman.

There are simply no words to express how much both of you are loved and missed.

Contents

Preface

On September 17, 1975, my life was forever changed. I was struck and killed by a bolt of lightning. I never saw it coming, and I died without a belief in any particular God or an opinion about the afterlife. I was twenty-five years old and wide open to just about anything. So, that's exactly what I encountered next—the wide, wide world of the afterlife! In the Kingdom of Heaven I was shown the glory of God as well as the magnificence of humankind. After spending twenty-eight minutes on the other side, I was sent back to spread the good word: *there is no such thing as death.*

In 1994, a publisher approached me in regards to publishing a book about my adventures in Heaven. At first I saw no reason to do that since I couldn't understand why anyone would want to read it. I wasn't a celebrity or anybody in the public eye. I was just a simple country boy with a great after-dinner story to tell; I hardly thought it had the makings of a best-selling book. Well, did I ever turn out to be wrong! With the tremendous help of my coauthor, Paul Perry, *Saved by the Light* spent a breathtaking *twenty-six weeks* on the *New York Times* best seller list. About the eighth or ninth week on the list, it dawned on me that people really were interested in learning more about life after death.

A year later, Harper Collins published our second book. This time, *At Peace in the Light* sold over six hundred thousand copies. The focus of this book was an in-depth look at the Boxes of Knowledge I had been shown during my first near-death experience. On that day in Heaven, over one hundred visions of the future were downloaded into my brain for safe keeping. I saw famine, war, and environmental destruction. I saw man-made disease, chip implantation, and manipulations leading to an astronomical rise in the price of gas. But I also saw the future of medicine moving into noninvasive light therapies—the kind that were to be implemented, for stress reduction, in the centers I was sent back to build.

It had been my plan to follow up quickly with a third book for my fan base, but I had a few problems along the way. First, I was stopped in my tracks by a major health issue that led to brain surgery. I confronted this head-on (pardon the pun) in 1997. Once I recovered, my entire mindset had shifted. The book I'd planned to write just wasn't appealing anymore. I'd had another near-death experience during brain surgery, and it contained information I was incredibly hesitant to share. Suffice it to say, it was far from the glory-filled extravaganza I'd enjoyed so much in 1975. However, the need to share my knowledge haunted me for years. As the twentieth century handed the baton to a brand new millennium, my wife, Kathryn, enthusiastically encouraged me to make known the secrets I'd learned in the light. I finally agreed to do it on one condition: if she would be my coauthor. I knew that with her skill as a writer and my flair for oratory, we could produce an excellent volume. And, in 2004, that's precisely what we did. *The Secrets of the Light: Seven Spiritual Strategies to Empower Your Life . . . Here and in the Hereafter* was self-published by our company, Heart Light Productions, with all of the proceeds donated to our non-profit organization, the Twilight Brigade, for training hospice volunteers.

Kathryn and I were pleased with the wonderful reception our book received. After selling out our first run, I decided to approach Harper One to see if, by chance, they might be interested in publishing it for me on the second round. Lo and behold, Editorial Director Michael Maudlin decided to take the bait. God love him. But, for me and Kathryn, the real work was just about to begin.

In going back over what we'd written three years earlier, we saw that there had been a lot of changes in the world and we had gleaned a vast amount of insight through experience. Most of all, there seems to be a new wave cresting over humanity's sea of perception exactly at the point of religion versus spirituality. In just under two thousand years, our holy outlook has gone from pagan to pious and back again to a more natural and loving spirituality. The faithful have journeyed from sacred sexuality in an oak grove to strict ecclesiastic control over every mundane detail of our lives. Then, in the 1960s, we brought back a renewed sense of free love and a more personal relationship with divinity. The lessons I learned in Heaven validate the need for a new understanding of our spiritual nature. As spiritual beings in the middle of a human experience, we have to reassess our eternal priorities. Forever is not nearly as long as you think it is—it's a whole lot longer. When we put the infinite into perspective, it becomes possible to grasp the amount of work that needs to be accomplished by each soul in the course of one lifetime. Since we are made in God's image, it will be necessary for all of us to continue to grow and expand, learn and mature, create and re-create ad infinitum.

This updated version of *Secrets of the Light* is designed to help in doing just that. It is a gift of love from Kathryn and me to the world. We sincerely hope the celestial lessons written upon these pages will create a deepening of love and understanding in the heart of eternal truth. The universe is systematically designed to assist us in the fulfillment of our personal desires. It just wants us to be happy. Based on my heavenly education, we offer practical instruction for the daily application of spiritual principles intended to empower and impassion your life—here and in the Hereafter.

PART I

There and Back Again

ONE

My Lively Dance with Death

The art of living well and the art of dying well are one.

—EPICURUS

The first time I died, it wasn't nearly as scary as I'd thought it would be. Yet, quite honestly, at age twenty-five, I had not actually spent a lot of time thinking about death—at least not my own. Nevertheless, between 1975 and 1997 I managed to die once and dance toe to toe with death through two more near-death experiences. I recounted much of my story in my first two books, *Saved by the Light* (1995) and *At Peace in the Light* (1996). But in this book I simply want to tell enough of the story to get us to the real purpose of the book—how we can all live more fulfilling and less fearful lives.

Back in 1975 I was one cocky son of a gun who could anticipate a punch, fieldstrip a rifle, and fix a '57 Chevy with one arm tied behind my back. Beyond that, and an occasional night out with my girl, I didn't give a damn about much of anything else. In my hometown I was the bully everybody loved to hate. Moreover, that's exactly how I liked it. I believed I had created the perfect world for myself, until a lethal thunderstorm loomed within throwing distance of my home in South Carolina on the evening of September 17, 1975.

I'd been working out of town for several weeks. While I was catching up with a close friend on the telephone, a bolt of lightning made a direct hit right outside my house. Like a heat-seeking missile, it was swiftly drawn through an ungrounded telephone line. Within seconds, it carved a path of melted flesh and bone as it entered my head through the phone receiver. It all happened so fast, there was no time to react. Although I heard a deafening sound, like a high-speed locomotive rushing toward my ear, the lightning flashed inside my head before I could assemble my thoughts to put down the phone. Then it was too late; the lightning was hell-bent on having its way with me. Instantly, it lifted me off the bedroom floor and held me suspended in the air. The searing pain stunned my senses as the electricity burned its way through my entire body, engraving its fiery initials down the length of my spine. My body was burning from the inside out. It was beyond excruciating—truly torturous to the point of being incomprehensible. I could not think; I couldn't even scream.

As the lightning exited through my feet, I dropped to the bed below as if forcibly thrown, severely bending the frame in the process.

By that time my girlfriend, Sandy, heard the commotion and came running in from the kitchen to see what was going on. She gasped in horror at the sight before her. There I was, badly charred and broken, strewn lifeless across the bed. She took a deep breath and immediately started CPR on me. I remember watching her do this. I was perched just under the ceiling at the far side of the room, and although I was floating above the scene, it never occurred to me that I was actually dead.

You see, at that time, the depth of my spiritual perspective was truly limited, if not fatally flawed. I was raised in the Deep South and, as you probably know, everybody goes to Hell where I come from. As a survivor of that fundamentalist religious environment, I possessed no spiritual framework for understanding what I was about to encounter head-on. I had never heard of a near-death experience. Dr. Raymond Moody hadn't even coined the phrase yet. And I wouldn't have believed in it even if I had heard of it. I never gave the time of day to that kind of cosmic crap. For me, the world of the mystical didn't hold the slightest interest. But as of that moment in 1975, everything spiritual had my full and undivided attention. I guess you could say the lightning had permanently readjusted my attitude!

Before too long, from my vantage point just below the ceiling, I saw Tommy—the friend I'd been talking to on the phone—rush into the bedroom. Panicked by the abrupt way our conversation ended, he'd put down the phone and headed over to my house. Having been trained as a corpsman in the Navy, Tommy automatically took over the resuscitation effort while Sandy ran next door to call for an ambulance. I watched them with great curiosity and without the slightest sensation of physical discomfort. That crushing, burning torment that had riveted me just moments before was now completely gone.

As all this transpired, the vibrant kaleidoscope of living colors that emanated from Tommy and Sandy astounded me. In fact, as I glanced around the room everything appeared to be literally alive and vibrating with color. Even the wooden chest of drawers in the corner radiated a multihued energy. What an amazing observation for a redneck! I wish everyone could see what I saw on that night. It certainly changed the way I relate to animate, and even inanimate, life to this very day. I no longer take for granted the unique and beautiful spiritual life force flowing through every creation in the physical world. To witness how intricately everything was connected, interwoven, and related, at the deepest levels of a highly organized matrix of networked energy, was indeed an overwhelming and inescapable new reality for me. Yet, as inspiring as that discovery was, it would pale in comparison to what was yet to come. Before the night was over, I was destined to witness the breathtaking marvels of the heavenly realms.

But first, winded and flustered, Sandy ran back from the neighbors to let Tommy know the paramedics were on their way. She was anxious to resume her efforts to restart my heart. As luck would have it, Sandy had recently completed CPR training offered by her employer. Furthermore, her confidence in the effectiveness of this procedure was exponentially heightened by her desperation to get me breathing again. In a last frantic attempt to revive me, Sandy raised clasped hands over her head and hammered them down upon my chest with a powerful and painful thud. At the moment of impact, my eyes sprang open and I drew in a labored breath. Instantly, I felt as if my body was being squeezed into a paralyzing straightjacket, one that had been soaked in acid. This agonizing state lasted only a couple of minutes before I started to convulse violently. Within seconds, I lifted out of my body again.

Just in time to intercede between Sandy and complete hysteria, the paramedics rushed into my house without wasting the time to knock. They began working on me with feverish precision. Before long I was loaded into the back of the ambulance with Sandy by my side. Behind the ambulance, Tommy drove his car the short distance to the hospital. I sat beside my body, opposite Sandy. The paramedic put the paddles to my chest several times but finally gave up. He believed I was beyond help. Yet he told Sandy we'd be at the emergency room soon, and they'd continue to work on me there. She wept quietly, her face in her hands. I tried to comfort her, to tell her I was there beside her, but she couldn't hear me. I recall studying my face and body very carefully. I remember being disappointed by my appearance. I'd always prided myself on being such a handsome stud, but tonight I literally looked like death warmed over. It's funny how pride and ego are the very last human vanities to go.

When the ambulance screeched to a halt at the hospital doors, several people came running out to meet it. Sandy and Tommy were escorted to a waiting room, and my body was speedily wheeled into the emergency room just in time to be pronounced DOA. It was a hectic Friday night at the hospital since the severe thunderstorm had wreaked havoc on the town. The examining room I occupied was needed for living patients, so my body was covered with a sheet and stored in any empty room until an orderly could be found to take me to the morgue. This definitely was not turning out to be my lucky day.

But I wasn't there for any of that. Instead of going into the emergency room with the gurney, I found myself enveloped by the shimmering blue/gray vastness of a whirling tunnel. Next, I was transported through space, feet first, as though I was lying on an invisible conveyor belt. Initially, all I experienced was deep silence, but soon I could detect the faint sound of chimes carried on the wind from a distance. Ever so gently, I felt my body vibrating in response to each tinkling tone. Still, the spiraling space continued to rhythmically coil round and round me.

At this point I saw a light at what appeared to be the end of this swirling vortex. The light emanated an incredibly brilliant and captivating glow. At first, I was afraid to look at it—I thought it would burn my eyes.

Yet I could not help myself. I felt irresistibly drawn to gaze into it. Surrendering to the temptation, I suddenly found myself standing inside the light. In fact, the light and I were one. I felt safe and complete, perhaps for the first time in my life. The light was alive. It infused me with its warmth and cradled me in what I sensed to be a sanctuary of all-consuming love and acceptance. The experience was beyond exhilarating.

The sensations verged on orgasmic, yet my thoughts were spinning a mile a minute. Where was I? What was this place? Could this all be a delicious dream about to end? Glancing down, the outline of my hand grabbed my attention. It looked transparent, and yet, it sparkled in little rivers of changing color. Looking at my body, it too pulsated with the same rainbows I had seen around my friends as they had worked so hard to revive me. I was taken by surprise as I felt the presence of someone drawing near. Through a thick yet translucent mist of iridescent silver, a Being of extraordinary light materialized before my eyes. Its energy felt nurturing, kind, reassuring, and oddly familiar. In searching for the perfect word to describe how I felt in this Being's presence, only the word *precious* will suffice. As it seemed to float toward me, I felt increasingly dearer; I became more treasured and cherished, the closer it got. However, the Being of Light granted me no time to indulge this new and unusual way of viewing myself.

Instead, the Being promptly proceeded to turn my attention in the directions of other souls, which I hadn't realized were moving all around me. As I focused on them, I could discern that they were occupying differing levels of energy and vibration. Those who appeared to exist below me were vibrating at a much slower rate, and focusing on them caused my vibration to slow uncomfortably. Those above me were lighter in density and emitted a higher frequency than my own, yet by looking at them I began to increase my own vibration. I wasn't sure what to make of this or why the Being wanted me to be aware of it, so naturally I turned to the Being of Light for understanding. As I did, the Being scooped me up in what felt like a huge embrace, one that launched me on the journey of my lifetime. This fantastic voyage blasted me into my past, beginning with the day I was born. From that point forward, I was shown the highlights from every year of my life, right up to the moment the lightning fried my body and claimed my soul.

I have since coined a phrase to describe this comprehensive afterlife spectacle. I call it the *panoramic life review.* Why? Because I saw my life in a 360-degree panorama, and it was skillfully produced and directed to remind me of my first twenty-five years of less-than-righteous behavior. No kidding. I was mortified as I watched the movie of my life unfold. I'd hurt so many people and acted out in so many cruel and ugly ways. From the kids I teased in the schoolyard to the enemies whose lives I had targeted for my country, I had lived a life that was harsh and violent.

To this very day, I consider the panoramic life review to be the pinnacle point of my time spent in Heaven. During this all-inclusive recapitulation of my entire history, every single thing I had ever thought, said, and done was displayed for me in holographic form. As the scenes played out (and often ended painfully), I was not only myself, but I became every person I had interacted with along the way. I experienced exactly what they felt as I relived their encounter with me. More often than I cared for, I became the frightened victim I had made of others. Viewing my actions in the way others had viewed them now evoked tremendous shame and guilt within me. Very quickly, I wished I had spent my life being a kinder, more loving person.

By the time my life review ended, I was aware of another vitally important fact: the universe is systemically designed to assist us in the fulfillment of our personal desires. However, there is a hook—and a sizable one at that. You see, within the system of universal consciousness, no mechanism exists for the discernment of what we call right and wrong or any sinful act. Simply put, the universe does not recognize the difference between light and dark or good and evil. Therefore, *we* must. It is up to us, for within us resides this innate knowing. We have come to this Earth for the express purpose of learning to master the proper execution of our preexisting spiritual wisdom. We will act as our own judge and jury. Take it from me, this is a far more eye-opening point of view than any religious perspective we might now maintain. And in the end, we alone will hold ourselves accountable for our every thought, work, and deed as well as the resulting ramifications in the lives of those we touch.

I felt emotionally drained when the review came to a close. Gradually, I began to experience a sense of increasing clarity and lightheartedness. It was as though the review had not only shown me the error of my ways, but it had also cleansed my soul of shame and sadness. It seemed as though reliving and acknowledging my dreadful mistakes signified my spiritual redemption—at least for the time being.

What I did not know was that I would be going back to the awful life I had just reviewed, and I was going to have to face the consequences for every miserable act of cruelty and thoughtlessness I'd perpetrated. For truly, each choice we make in life creates a consequence we will eventually have to face. The Good Book says, "For whatsoever a man soweth, that shall he also reap." My mother used to put it this way, "What goes around, comes around." And she was usually coming around the corner chasing after me, swinging the kitchen broom at the side of my head! Mother was real big on discipline, if you know what I mean. Anyway, back at the life review screening room, the Being of Light let me know it was time to go. I would like you to understand that the Being never spoke to me in words. Yet, our exchanges were so much more than what we think of as mental telepathy. Inexplicably, the Being's thoughts and movements created an intimate energy capable of suffusing me with a silent understanding that permeated my being to the core.

In the deep purple distance, I could see the silhouette of a city of light etched upon a mountainside. The Being and I lifted, as if remotely controlled and drawn toward this magnificent sight. As we came closer, the lights grew brighter and glistened exquisitely. The entire landscape filled me with awe. In the greatest splendor before me there appeared Crystal Cities of breathtaking beauty. As we arrived at the entrance to one of the cities, I beheld row after row of gothic-like cathedrals in varying shapes and sizes. I had always loved the study of period architecture; therefore, this was a rare and mesmerizing feast for my eyes. Unfortunately, there was no time to tarry. The Being led me through an annex and into a great hall. I could only describe it as a hall of knowledge much like I envision the Library of Alexandria must have been before the caliph Omar burned its priceless treasures to heat bathwater for his Arab troops. This remarkable hall radiated a warm, golden glow that made everything in the room feel like liquid love.

We were graciously welcomed into this celestial chamber of wisdom by twelve additional Beings of Light, standing behind a podium made of crystal resembling clear quartz. At that point, I could no longer sense the energetic presence of my original escort. It just seemed to vanish, or perhaps it merged with another of the Beings. However, at that instant we were joined by a thirteenth Being. All of these new Beings were larger and far more majestic than my original host. They commanded great respect simply by virtue of their strength and stature. I remember feeling that each one of them represented a different spiritual virtue such as hope, courage, faith, or compassion. Although they did not appear to be either male or female, I sensed they were both. The thirteenth Being, the last one to appear, seemed to be the chairperson of the board, for it exuded great wisdom.

I had no idea why I was being granted this audience with them, yet I felt so safe and calm in their presence. Within this sacred space, I found myself immersed in the intimate vastness of our original home—the place we call Heaven. Within this sanctuary of universal holiness, I was swaddled in the rapture of the divine. I could not help feeling overwhelmed. In a burst of euphoria, my mind, heart, and soul opened simultaneously to complete cosmic awareness. In that instant, I knew everything there was to know. I was one with all creation, and it was one with me.

And it shall be the same for you, no matter how spiritually evolved you may be. For you see, nothing on Earth can possibly prepare you. Our humanness simply cannot conceptualize the breathtaking sense of wholeness that reunites with our spirit in the experience we refer to as death. My transition from this life to the afterlife was a mind-altering explosion of universal oneness that encapsulated my totality. In this process of spiritual conversion, I sensed the meticulous perfection and orderly unfolding of the entire universal plan. More than that, I glimpsed the heart of eternity, and then I knew unequivocally there is no such as thing as death. Life, like love, is forever. Together, both life and love constantly seek to expand their expression throughout the universe.

The Beings generously gave me those few moments to glory in my newfound thoughts of cosmic perfection before they, one by one, stepped up to face me directly. From their mind to mine, they sent me black boxes—one at a time—each about the size of a videocassette. The boxes literally passed through my head and, as they did, I was shown what I can only refer to as videoclips depicting visions of the future world. Some were thrilling; many more were bone chilling.

I saw the fall of the Iron Curtain ending the rule of Communist Russia. I witnessed the bombs falling at the onset of the first and second Gulf Wars, along with protests against the wars by women dressed in black. I was shown American military troops lining the U.S./Mexican border as well as the ultimate declaration of the Republic of North America uniting Canada, the United States, and Mexico as one country. I saw medicine moving into the astonishingly effective complementary and alternative modalities, such as a myriad of phototherapies. I saw cancer, and other life-threatening diseases, eradicated with healing wands made of light. I also saw the invention and implementation of a nanochip with the potential to be used for good or for evil. It was wonderful to understand its implications in the finding of lost children or the tracking of missing military personnel in a time of war. But I was just as quickly shown that it could be used to monitor the activities of innocent citizens and ultimately to detonate a deadly virus into the body of an unsuspecting victim that any government wanted eliminated. More than that, I saw that the chip could easily be implanted in elderly as a traceless form of euthanasia. This was an eye-opening and bone-chilling prospect.

I continued to be infused with these vivid scenes from the future of our world until I'd absorbed well over one hundred scenarios. Sensing my anxiety, and in the hope of assuaging my fear, the Beings told me that our future is not written in stone. We still have the chance to change the outcome if only we return our collective consciousness to the reality of love. The Beings then shared what my personal mission was to be. They told me I was to establish spiritualistic capitalism on Earth. They said spiritualistic capitalism was needed because people had to realize that their security is not to be found in governments, institutions, or religions. All three of these are necessary components of human life, but they were not meant to be relied upon completely. Rather than living in a society ruled by godless capitalism, the Beings presented the idea of each one of us finding something we love to do and then using that talent or gift to serve the world while also making an income. That is the definition of spiritualistic capitalism.

The Beings emphasized that our highest human potential can be reached only when we remember who we are as great, powerful, and mighty spiritual beings. In Heaven, we are considered heroes just because we were valiant enough to leap fearlessly into the great adventure called life. Leaving the spiritual kingdom in order to be born in the physical domain requires immense fortitude and unshakable faith. For that very reason, the Beings are always standing by to assist us through any challenge. All we have to do is ask and they are ready to show up for us with inspiration, motivation, and insight. On rare occasions they will actually make their presence known through circumstances we call divine intervention.

During my last few minutes in Heaven, the Beings gave me one final task: create centers for stress relief. They indicated that part of the reason many people were not living in love and harmony was the direct result of too much stress. Stress carries a heavy energy that attracts negativity and fear. In turn, this slows our chakras, drains our spirit, and disconnects us from our divinity. By relieving the stress in our daily routines, more light and laughter would be allowed to filter through our lives, making us all healthier, happier beings. The centers were an eight-step program, by design. I then saw a vision of the seven rooms in the center:

- a therapy room
- a massage room
- a sensory-deprivation chamber
- a bio-feedback room
- a psychic reading room
- a room for a *Klini*—a specially created bed intended to induce the deepest levels of relaxation
- a reflection chamber

In the eighth step, the client is taken back to the Klini for more relaxation through out-of-body travel. In this step, people can actually visit the spiritual realm to connect with departed loved ones. Each step of the center is to create a sense of security through deep relaxation. This empowers people by reconnecting them to their true inner self and pure spiritual nature.

Once these assignments were given to me, my heavenly hosts wasted no time in telling me that I had to return to my body. I couldn't believe it. Were they kidding me? As it turned out, no, they weren't! I was dazed, but there was no way they were getting me to go back to the pain and agony I'd left behind—certainly not after being in this sacred space of unconditional love and tranquility. They could forget it. I didn't care what they wanted. I'd taken on crowds a lot tougher than the likes of them in my bar brawling days! Well, guess what? Without a single punch thrown, I lost that fight.

The next thing I knew, I was in the hospital hallway floating above my dead body without a clue as to how to get back into that thing. It was stiff and cold with no visible entrance sign or chimney to shimmy down. What the hell was I supposed to do? Dumbfounded, I hovered there thinking that I'd had about enough adventure and intrigue for one day. This situation was working my last nerve. Not a second after that thought passed through my mind, I found myself looking up at the white sheet shrouding my body. But I couldn't move. I was completely paralyzed and once again burning from the inside out. Where were a couple of those Beings of Light when I really needed them? Panic rushed through me. Then I took a breath and let out a sigh of deep resignation.

Dear God, please help me. . .

TWO

Living Between the Worlds

There are two ways to live your life.
One is as though nothing is a miracle.
The other is as though everything is a miracle.

—ALBERT EINSTEIN

While I was dead and roaming Heaven, I felt more alive than I ever had among the living. Then I found myself on the stainless-steel gurney under a sheet and dressed only in a crumpled toe tag. For sure, I was not happy about being back on this side of life. But I was back and, like it or not, I was going to have to deal with it. Yet to add to my dismay, I discovered I could not move a single muscle except for those in my face. So once it finally dawned on me that I was breathing again, I started blowing up at the sheet covering me. Hopefully someone would take notice of the corpse that could breathe. My prayers were soon answered. It was my faithful friend, Tommy, who turned out to be that someone who was paying attention. As you might imagine, sheer hysteria ensued when everybody realized I'd come back from the dead. The hospital staff went to jumping all around, and my family had to decide whether or not they were happy to hear the news! I'm only kidding—I think. Nevertheless, I had no choice but to take the first step on what would prove to be my very long and tumultuous road to recovery.

I spent approximately six days in the constant care of my local hospital. Two of those days, I spent in a semicoma. When I awakened, I was still completely paralyzed for the next four days. On the seventh day, I managed to lift my hand to scratch the tip of my nose. After that, my family tentatively agreed with the physician to take me home. They wanted to allow me to die (again) in peace, surrounded by my loved ones. However, I did not die. Much to my family's surprise, I kept getting better. Much to my surprise, I found myself living life between the worlds. By day, I struggled to remember my life the way it was before the lightning. Who were all these people coming in and out of my house all the time? The names and faces of my closest friends and family were often unfamiliar to me. In turn, it created days filled with emotional turmoil for everyone involved. Frustration typically consumed me when I couldn't recognize someone I was supposed to know. And then my heart would break when I'd sense the deep disappointment evoked by my amnesia. In the early weeks of my recovery, too many days felt like an unpredictable ride through the Little Shop of Horrors.

I could never imagine what distressing surprise lay around the next turn. Therefore, with the coming of the night, I treasured the promise of sleep. Automatically, I was transported back to the Hall of Knowledge where I would be cradled within the calm heart of the Crystal Cities. There I would be tutored in the mysteries of the universe on every possible subject from astronomy to zoology. My nightly travels to the celestial realms were what I began to live for. I craved these nocturnal dispensations; I couldn't absorb enough knowledge. On a nightly basis, my soul's appetite for this cosmic education grew more and more insatiable.

Slowly, the days passed and I continued to grow stronger. When I could manage it physically, I'd sit at my desk and write down every detail of what I was learning in Heaven. I filled piles of spiral notebooks with the facts and formulas given to me on the other side. Some of the subjects were taught through a kind of osmosis. I'd just sit there and feel myself absorbing the curriculum, much like a computer downloading files. The other courses were taught by actual Beings. And I do mean beings, because they definitely weren't all humanoids. Some of my instructors were inter-dimensional, with no definable outline or form. Some were from higher dimensions, otherworldly in appearance, similar to what some describe as extraterrestrials. They all had one thing in common, though—they were all beyond what we call genius. I was completely enthralled by their tutorials chock full of priceless wisdom, and my every question was answered before I could even ask.

Gradually, though not fully aware of the transition, I no longer had to wait for nightfall to bring my bliss. The Hall of Knowledge started showing up during the day in my own living room. Unexpectedly and right before my eyes, the room would morph into my dream-time study hall. One moment I would be looking out the window from my favorite rocking chair, and the next I'd be gazing out over the purple hillside from high atop the Crystal Cities. Initially, I was astonished by this time/space travel. To be completely honest, I was even a little scared. But over time, I got used to this regularly scheduled phenomenon. You see, in light of the roller coaster of physical and emotional challenges confronting me during my waking hours, I resigned myself to these spiritual journeys as just another part of my brand-new life.

Not quite as easy to accept, however, were my newly acquired psychic abilities. Since the day I awoke from the coma, I was able to hear the thoughts of the people around me. Of course, in those early days I didn't have the strength or concentration to do more than acknowledge it. Over the following weeks, however, I started to pay more attention to listening to what people were thinking, because it was often the opposite of what they were saying. I found this very curious—like the time Mother came to see me with her cousin, Ellie. Mother looked at me smiling and said, "Danny, look at you! You're looking so much better today." But what I heard her thinking was how the sight of me made her queasy. I came to understand what little truth people tend to express under duress, whether they're trying to mask fear, doubt, worry, or grief. Obviously, Mother feared for my well-being and she did not want to worry me with her angst. I appreciated her concern about half the time; the other half, I still couldn't remember who she was! But that never kept her from checking on me every day. My dad never failed to be there when I needed him either.

Overall, it took me two difficult years to recapture full mastery of my body. Since I'd always been healthy, strong, and agile, these were physical attributes I'd taken for granted—until I was struck by lightning. I promised myself never again to eat a meal without saying the blessing, to ignore the beauty of a sunrise, or to forget to say thank you for the smallest kindness extended by another person. In our lives on this side of the veil, we have no way of knowing which day on Earth might be our last, so we are wise to truly appreciate each moment as a precious and irreplaceable gift.

THREE

So Far from Home

Security is mostly a superstition. It does not exist in nature, nor do the children of men as a whole experience it. Avoiding danger is no safer in the long run than outright exposure. Life is either a daring adventure, or nothing.

—HELEN KELLER

As I eventually grew strong enough to venture from my bed to the kitchen, and from the kitchen to my rocking chair, my mind longed to venture toward Heaven. There remained so many haunting and unanswered questions still weighing heavily on my heart. It was simply impossible for me to reconcile what I'd seen in Heaven with what I was witnessing back here on Earth. I felt lost, like a child stranded far away from home and everything familiar. Day after day my real home—Heaven—seemed to be moving farther away from me. I grew increasingly depressed, for I longed to hear the sacred sounds and to see the breathtaking colors that one can only behold in Heaven. I missed being dead!

Incredible as that might sound, it was true. I'd traveled to the other side of the veil, and I knew there is no such thing as death. What we refer to as being dead was the most breathtakingly beautiful experience I'd ever had. The afterlife is real, more real and more utterly magnificent than anything we could ever dream of on this side of the veil. Therefore, one nagging thought in my soul constantly begged for an explanation: Since our everlasting home is so splendid and our souls are immortal, why do we have to come to Earth at all? Why do we have to endure spiritual suffering, mental poverty, physical sickness, and emotional loss when our eternal reality bears no likeness to anything like that? And where on Earth could I possibly turn to find someone who would possess the wisdom of the Beings of Light I'd just left behind in Heaven? Who were those Beings, anyway? Were they angels, archangels, ascended masters, what?

No doubt, I was a spiritual being wading my way through the ultimate human conundrum. As much as I loved the Carolinas, in 1975 they were the last place to search for advanced spiritual truth. Nevertheless, I had become obsessed with finding the answer to my personal riddle. Thus my quest began, and as a heartbreaking consequence, one of my closest personal relationships began to erode. Though she honestly tried, Sandy just could not grasp why I was so possessed with this insane quest for truth. I was driving her mad with my incessant chatter about the Beings of Light and the Boxes of Knowledge. She'd done everything in her power to help me recuperate physically, but she felt completely incapable of facilitating my spiritual recovery. All she wanted was for me to be normal again. However, after what I'd seen in Heaven, normal now appeared to be grossly overrated.

The truth is I might have genuinely gone more than a little crazy after spending all those months crippled and at the mercy of my pain. I sincerely believed my mind was the only thing still working properly, while everyone else whispered behind my back that I'd lost it completely. Before long the unrelenting strain this put on my relationship with Sandy proved to be too much, and we came to the mutual conclusion it was no longer possible to negotiate our emotional contract. One dark day she moved out with my heart packed neatly away in the bottom of her suitcase.

Losing my relationship with Sandy was devastating. It created a void in me that expanded the existing empty space in my soul—space longing to be filled with truth. But it had to be truth in the form of spiritual comfort. I soon started a brand new relationship, this time with the Holy Bible. Night after night, I scoured every verse for even the tiniest morsel of hope and enlightenment. In Heaven, I had been bathed in love and splendor; I'd experienced unbelievable joy. Now I hungered for a message in the Good Book. Was there even one chapter or verse that could explain why I'd been forced to leave my heavenly home? Could a scripture exist to help me find inner peace on Earth after witnessing the beautiful realm of the angels? If there was, nothing would keep me from finding it.

Sunday mornings soon brought me an even more entertaining slant on my expedition through Christianity. I found Reverend John Scott, Tammy Faye and Jim Bakker, Oral Roberts, Kenneth Copeland, and Herbert W. Armstrong to be a trip. I would sit in front of the radio or television, positively mesmerized by the lyrical sermons of these savants of the Sabbath. Looking back, perhaps everyone else was right. Maybe I was going mad, crazed by the prospect of being stuck in the physical world after my eyes had seen the celestial wonders. I yearned to be a part of the spiritual realms once again. An obsession for finding the truth had a hold on me, and it was not going to let go until I found what I was looking for. I knew I was being driven to search for the divine meaning of life, and it had become an obsession to be reckoned with.

After many months of Bible study, my spirit grew restless. Though the Bible's consecrated prose repeatedly offered me glimmers of hope, it seemed that the more I read and the more I heard, the less I understood. Then, without notice, my route to cosmic understanding took a detour. I felt called to investigate the spiritualist movement that lasted between 1840 and 1920. Great books by authors such as Sir Arthur Conan Doyle, Madame Helene Blavatsky, Edgar Cayce, and Evangeline Adams, to name just a handful, became my spiritual mainstay for the next intriguing phase of my journey. I gleaned an immense amount of inspiration from them all. These turn-of-the-twentieth-century spiritualists were truly enlightened seekers, sages, and channels who offered wondrous bits and pieces of the answers I sought.

Yet again, my mind craved more substance. I had honestly experienced the bliss and grandeur of supreme divinity. These fine folks possessed only a limited, though educated, perspective on the subject. No matter how eloquently they expressed their views, they had not seen what I had. They had never gone where I had been. Stymied and at the end of my rope, I began to lose my hold on hope. I truly feared I would never again find my peace of mind. I felt like an innocent man given a life sentence. My soul had soared through the heavens, I had beheld the awe-inspiring reality of the world beyond, and now I was confined to this broken body and tortured mind. I needed God to give me another call because I had no idea what he was thinking.

I waited, but I didn't receive any divine communication, so I forged ahead—or rather—into the past. Determined to discover the secrets of life, I deftly combed ancient texts of the Sumerians, the Maya, the Babylonians, and countless other civilizations. I refused to relinquish my hope of finding what I needed most—absolute, undeniable confirmation of the astounding spectacle I'd beheld, Heaven. The ancient Egyptians left behind a hieroglyphic library filled with everything they knew about the glorious mystery called death. Like the Maya, they had successfully, and ornately, erected an entire culture around the rites of the last human passage. Surely someone among them had pierced the veil and glimpsed this unknown as I had. The remaining libraries of antiquity had to bear a lost or secret text devoted to the afterworld. Without a doubt, I believed there must be one last scroll or tablet in which my soul could find verifiable solace. But alas, this search through the succession of our ancestors was also destined to deny me absolute fulfillment. Finally, I was starting to get the big picture: I could trust myself and have faith in the information I had already received.

For a fleeting period of time, my enthusiasm was reignited by the writings of Zecharia Sitchin. The author's first book, *The 12th Planet*, was filled with fascinating tall tales of the Annunaki (in reference to the giants, or Nephilim, of the Old Testament) that kept me captivated for days on end. After devouring that book, my appetite was insatiable. I plucked the meat from the bones in search of the wisdom of the Gnostic Gospels as translated from the Nag Hammadi codices. That led me to the secret teachings of the Essenes. This collection of ancient knowledge was more than enough to fill my mind and heart with pure joy. I was learning so much about the origins of modern religion and spirituality, I felt as though my life was being purified from the inside out. Yet, I still had not stumbled upon the capstone, that finishing piece to my edifice of understanding.

If I did not come to terms with my personal truth, I would most assuredly drive myself, and everyone who loved me, slowly insane.

So finally, I began to pray. I prayed incessantly. I possessed only one human aspiration at this point: to discover just one other person who could validate what I'd seen during those twenty-eight minutes in Heaven. Let me tell you, prayer is a many-splendored thing. I know this because God sent, and right on time, Dr. Raymond Moody. I believe this timid, soft-spoken, yet understatedly remarkable man was born to change my life forever.

FOUR

What You Seek, Seeks You

No man or woman of the humblest sort can really be strong, gentle and good, without the world being better for it, without somebody being helped and comforted by the very existence of that goodness.

—Alan Alda

I have to tell you, I never expected to be met on the other side by a bright, beautiful Being of Light. See, I'd been told, by teachers and kinfolk alike, for as long as I could remember, that I was so bad I was definitely going straight to the bowels of Hell. And I believed it! I knew what I had put Mother and Dad through and unfortunately for them, Ritalin wasn't around back then. So you bet, as I approached the end of the tunnel, I was fully prepared to confront the huge, vicious demons I imagined would be standing guard at Hell's grand entrance. I figured that maybe even Satan himself would be there to welcome me home. However, much to my amazement, no great big demons or little-bitty devils rushed up to meet me in the afterlife. No shiny red, cloven-hoofed, pointy tailed meanies could be seen pulling back the tarnished gates of Hades. Instead, my experience proved to be just the opposite.

In the light, I was engulfed in celestial love and serenity. I felt utterly cherished, as if I were the most treasured being in the universe. Now I yearned to feel that way again, to return to that realm of bliss. Although I labored to recover my strength from the physical damage caused by the lightning, the experience had left me emotionally drained. I felt isolated. Life had become so confusing, heartbreaking, and hypocritical—I just wanted to go home again, home to the place I had witnessed the true glory of the Divine Spirit, where there was no pain or confusion, no grief or violence. My inner spirit was growing weak. I needed to understand why I was still alive, why I'd been sent back to live in such pain and bewilderment. I wasn't sure I could hold on much longer. And then I got a love note from Heaven. While browsing through the local paper one rainy afternoon, I came across an advertisement that thrilled my heart and revived my spirit. Raymond A. Moody Jr., M.D., was scheduled to present a lecture titled, "The Near-Death Experience" at the University of South Carolina. Good God! I felt like I had just won the lottery. Hope was alive and well for a southern boy that day!

Finally, I was going to meet an expert who—I desperately wanted to believe—would be able to explain where I'd gone when I died. Maybe this doctor could actually help me put the pieces of my life back together. The ad said this brilliant young scholar had been doing research on the near-death experience for over a decade. He had interviewed more than fifty people who, just like me, had died and come back to tell a tale of traveling down a tunnel and into the light. I sensed that Dr. Moody and I were going to become good friends and if I was really lucky, maybe this new friendship would mean my dark days were about to come to an end. The next two weeks couldn't go by fast enough.

Throughout Dr. Moody's fascinating presentation, I sat awe-struck and spellbound. Everything he touched on in his lecture—and so much more—had happened to me. During his talk, it was hard for me to sit still without giving a spontaneous shout to Heaven! More importantly, I just couldn't wait to ask him a couple of important questions.

After waiting for my turn in a line of over fifty people, I finally got the chance to speak one-on-one with Dr. Moody. Astoundingly, his excitement equaled my own. I told him my name, and he actually recognized it from the article that had been published in the local newspaper days after I was struck by lightning. Since that time, Dr. Moody had been hoping to find and interview me. Now here we stood, face-to-face, both of us in search of answers to questions that were burning like wildfire in our souls. From our very first meeting, we were in constant communication.

It was an unbelievable dream come true, for me, to spend time talking with Raymond on nearly a daily basis. Unlike my other friends, and all my family, he encouraged me to recount my near-death experience over and over again. Everybody else I knew prayed for me to go mute! They were all so tired of my endless ranting. They simply could not relate to my story since none of them had any grasp of what I'd been through. On the other hand, Raymond absorbed my every word just like a thirsty sponge. He put my mind to rest and assured me I was not crazy. He could continuously fire deeply probing questions, questions that helped me to dive into my psyche from fresh and expanded viewpoints. As a result, I could explore completely different perspectives of the near-death phenomenon. With Raymond's able assistance, my quest wasn't just personal anymore. The anger I'd held on to with such self-righteousness was slowly starting to fade, and I was happily relinquishing my victim mentality in favor of a more historical and encompassing outlook. Finally, I could see how what happened to me could prove beneficial for the greater good of all, because people cannot truly live up to their greatest potential if they are wasting their energy on the fear of death. With Raymond's vast knowledge of psychology, history, and philosophy, he guided me to move outside my own story, to see this experience as universal and archetypal—representing the classic themes of infinity and immortality.

Through it all, Raymond kept me laughing, often until I cried. He had a good sense of humor and a dazzling intellect. Moreover, he was as determined as I was to leave no stone unturned when it came to researching the historical myths and modern tales of life after life. We were well matched as an investigative team in search of Heaven in all its mystery and magnificence.

One thing we had in common was rocking chairs. Raymond and I both loved them, and they hosted our best thinking and most intriguing conversations. Over time, our rocking chair discourses evolved into mesmerizing public presentations before sold-out audiences. It soon became apparent to us that we were not the only ones who wanted to know more about life in the Hereafter. So we decided to pack our clothes into the plastic garbage bags we used as suitcases and take our heavenly memorandum from city to city. I was beginning to enjoy my life again. I felt I had a purpose. It was deeply gratifying to be able to help people end their fear of death and come to terms with the fact there is a life after this one.

Sometimes after our presentations, I would be stopped by someone in deep emotional trauma—like a mother who was grieving over the death of her child or a suicidal husband finding it impossible to accept the loss of his beloved wife. Often I would be able to locate their loved ones on the other side and bring back loving words of reassurance for them. To be able to offer them that modicum of comfort became a huge victory in my life. In countless encounters, I was placed in the uncomfortable position of being asked to redeem people from their lives of quiet desperation. Indoctrinated by their religions to believe themselves worthless, helpless, and born in sin, these fine folks turned to me for salvation. I went from dead guy to savior (and that's when I discovered the many pleasures of Grand Marnier)!

Night after night, I would drop into bed exhausted, but happy, because I knew I had made a difference in someone's life that day. I had been shown in Heaven how important it is to initiate random acts of kindness in the lives of others, but this was the first time since I was struck by lightning that I was able to do so. I had been shown that whenever I took the time to share love and compassion with someone else, the whole world was touched by kindness. Being an inspiration to others was truly wonderful. Even so, after a while, I was ready to build a real life for myself, and I wanted it to consist of more than just talking about death and communicating with those on the other side. People around me were starting to act like I was some kind of messiah, and I was starting to buy into that notion a bit myself. Thus, the time had come to reinvent Dannion. At the very least, I had to find myself again. Obviously, the way I had been living my life was not working (and the price of Grand Marnier doubled) and so it was time to make a few changes.

This led to a series of businesses—the only things in common between them being long workdays and not taking care of myself. After many months of working eighteen hours a day, my health would hand me another near-death experience.

FIVE

Let's Synchronize Our Watches

There are no mistakes, no coincidences. All events are blessings given to us to learn from.

—ELISABETH KÜBLER-ROSS

To relax I rebuild cars. After searching for months, I finally found myself a 1958 Borgward Isabella convertible. What an outstanding automobile. On one Saturday afternoon, I was having a particularly difficult time removing a manifold. Try as I might, I just couldn't get this stubborn piece out. In the middle of my continued efforts, my forearm made contact with the manifold, and I suffered a nasty third-degree burn. That sucker hurt so badly I had to suspend my work on the car for the rest of that day. Instinctively, I put a little antibiotic cream on it, and then I sat down to enjoy a glass of wine or two. The next day, I went right back at it, and this time I sliced open my left hand while putting that same troublesome manifold back into place.

For days afterward, I sensed myself growing weaker. The Hong Kong flu was having a spring fling making its way around the world, so I thought I'd caught the bug, but I was finding it impossible to regain my strength and vitality. Finally, I figured I had to make a quick trip to the local emergency room.

"Mr. Brinkley, do you want me to tell you the truth?" a young doctor who looked amazingly like Doogie Howser asked me.

I answered, "Of course."

"Okay, you have a touch of pneumonia in your left lung. But more than that, I have to tell you, you're in heart failure. If we don't get you under immediate medical attention, you'll have only about forty-five minutes to live."

I looked at the young physician and asked, "Forty-five minutes? Damn, Doc, you reckon I ought to sit down? Let's synchronize our watches! Forty-five minutes and I'm out of here." That was the last thing I remember.

Two days later, I came around. When I awoke completely, my dad as well as other friends and family members were there. My dad told me how everyone agreed that I needed to have open-heart surgery if I wanted to live. It turned out that during the two days I slept, Dad had given the doctors permission to perform a coronary angiogram on me. They concluded that the cut on my hand and the burn on my arm had together created the perfect vehicle to transmit staph into my blood stream. The staph had found its way to my heart and proceeded to dine on my aortic valve. By the time I arrived at the emergency room, I was drowning in my own blood.

Dad insisted that I give my consent, allowing the open-heart surgery to take place immediately. I couldn't help laughing to myself at the time. How could they all think I would want to stay alive in this weak and vulnerable condition? I'd already been to Heaven—had they all forgotten? I had seen the wonder, the heart-stopping magnificence of the Hereafter, so there was nothing about being alive in this body that I wanted to prolong. I was stumped. My family had to be kidding me!

But Dad refused to give up. He was determined not to stop until he found someone able to talk me into having my chest cut in half with a buzz saw so my heart could be removed in order for the doctors to install either a stainless steel valve or pig valve in it. Dad had everyone he thought could influence me come to my bedside in an attempt to convince me to have the surgery. In many ways, I was flattered and appreciative. I knew how much my dad loved me as well as how adamantly he longed for me to stay with him. He had lost Mother only five years earlier, and losing me too might have been more than he could handle.

As a final desperate resort, he put in a call to Raymond Moody. Dad asked him to please come visit me in Charleston. When Raymond arrived a day later, I must admit I was happy to see him. Raymond was a most cherished comrade who would always have a place in my life. He walked into my hospital room and quietly approached me with his signature calm demeanor. He started our conversation by asking me what had brought me to the hospital, and then he questioned me about my feelings surrounding the prospect of staying alive through open-heart surgery. Once that discussion was complete, Raymond launched a campaign to change my mind. He told me how our mutual and most significant work would never be complete without my presence. He played on my sympathies. Moreover, he bet everything on the fact that I'd feel tremendous guilt over not living up to the vows I'd made to the Beings of Light.

When all else failed to evoke the desired response, Raymond decided to call on the classic arts of shame and chastisement to sway me. He challenged my integrity by accusing me of not fulfilling my highest mission if I refused to have open-heart surgery. Raymond let me know, in no uncertain terms, that the future of the NDE research would suffer a mighty and perhaps irreparable blow should I decide to return to the other side at what he considered a premature point. In my weakened condition, I could only take so much of his intense, loving, and relentless coercion. To my own chagrin, I reluctantly granted permission to the doctors to perform the surgery. In my heart, I submitted for the sole purpose of finally bringing my heavenly assigned work to fruition on Earth.

The instant the tip of that surgical saw touched my chest, I catapulted out of my body. I rolled over, in spirit, to see myself lying on the operating table. I watched as the doctor skillfully wielded his scalpel through my skin and sawed open my sternum with his high-speed electrical tool. He slowly turned the wheels on the device that spread my chest wide open. I witnessed him taking my heart out of my chest cavity and placing it on a square silver plate. My heart continued to pump for another seven beats or so, and then, it abruptly halted. I was dead—again.

It impressed me that no one in the operating room thought much of what was happening to my body. The other attending physicians were engrossed in a lively conversation about their upcoming fishing trip and didn't appear to regard my body as a physical presence in the room. At this point, it all was becoming too much for me. I heard the familiar sound of distant chimes. I found myself being transported through the tunnel and moving toward the shimmering brilliance of that comforting and beautiful light. My second near-death experience was poised to commence, and I was looking forward to it.

My second celestial excursion was basically the same as the first time around. Just like the first time, not a single soul came to meet me in the vestibule of paradise. Not Satan, not my mama, not even one furry critter I had ever owned, took the opportunity to welcome my arrival. I guess that gives you a pretty good idea of the general effect I'd had on people, and pets, through the years. However, the magnificence of the Being of Light—the same tour guide I'd had in 1975—who took me through my subsequent panoramic life review was undiminished in its brilliance. After the life review was complete, one very different thing did indeed occur. An extraordinary sense of pride and accomplishment flooded through me. I certainly hadn't had that experience the first time. In the fourteen-year interval between near-death experiences, I had worked hard to change my ways. Now I was shown that I had been triumphant. I witnessed a considerable difference in my behavior—I was more compassionate and far more thoughtful than the person I'd watched in my earlier life review. In this new version, I was impressed with the way I automatically reacted to the world around me. In the old days, everything had been a calculated response. This new picture expressed a conscious, gentle, and deeply tender essence.

Simultaneously, at a cellular level, I absorbed the importance of true forgiveness. In my life I had undoubtedly viewed forgiveness as a foolish weakness, for I had been a man with a contaminated point of view. I'd seen the cruelty and corruption humanity thoughtlessly perpetuated on itself. My heart had been tainted as a result of all I'd witnessed. But during my second near-death experience, the Beings of Light showed me the great personal power available to each living soul when true forgiveness comes into play. I learned from my celestial teachers that I needed to forgive myself. This was most important. And then I needed to forgive everyone else who I felt had done something to harm me or the world in general.

Of all the advanced instruction I downloaded throughout my second expedition into Heaven, the astounding beauty inherent in the act of forgiveness stands alone as the greatest value and consequence—both here and in the Hereafter. Without forgiveness, there can be no ripening of the soul, no authentic way of measuring our spiritual evolution. Therefore, forgiveness is indisputably our passport to divine greatness while we're still engrossed in the experience of our humanity.

By refusing to forgive ourselves, or someone else, we inadvertently deny ourselves access to the highest spheres of happiness as well as the deepest levels of our essential spiritual nature. I honestly had to take some time to bask in the poignant appreciation of this celestial insight. For you see, I had really relished holding on to my grudges. Some of the best times I could remember were spent nursing my emotional wounds. Now, for the first time, I realized how harboring those grudges only bound my energies to the same people I resented. As a result, my life and theirs would be significantly intertwined until the day I made the decision to open my heart and forgive. I remember reading once that to forgive means to forget as though a thing never happened. Forgiveness truly takes great strength and mighty courage, yet the rewards are beyond imagining.

On that day in Heaven, I understood that we are on Earth for only one reason: to act as the living reflection and expression of divine love. As long as we live in constant awareness of love as our reality, we stay in alignment with divine order and remain anchored in the field of higher consciousness. Forgiveness is love in action. Through the sacred act of forgiveness, we exponentially expand our ability to reflect more love and joy in the world.

SIX

No More Excuses

If I can stop one heart from breaking,
I shall not live in vain;
If I can ease one life the aching,
Or cool one pain,
Or help one fainting robin
Unto his nest again,
I shall not live in vain.

—EMILY DICKINSON

Getting back on my feet after open-heart surgery was a much more painful process than recovering from the lightning strike. The doctors gave me nine months to live if I didn't have the recommended follow-up surgery. They wanted a second opportunity to strip a muscle out of my back, open up my chest, and wrap that back muscle around my heart with a pacemaker implanted. That was the only way I could live without a heart transplant. After this ghastly procedure, not only would I look and walk like Quasimodo, but my life would have to start all over again. I wanted no part of it!

Instead, I decided to suck it up. I was going to make the very best of every day I had left. If nine months was all the time I had, then I was determined to turn it into nine unforgettable months. A strict regimen of antibiotics and the audible ticking of the St. Jude's prosthetic heart valve reminded me of the fragility of my life. But they also served as recurring reminders of what I had yet to accomplish, and the extremely short time I had in which to do it. The Beings of Light had given me assignments I hadn't yet completed, and now time was of the essence.

In the past, I'd whined and pouted, even dragged my feet in procrastination. But now there was no time left for my petty excuses. I called Raymond Moody. He was the one who coerced me into the heart surgery with his pitiful plea for my help in his NDE research. Okay, fair is fair, right? I had submitted to the surgery for him; now I needed him to get the research underway for me.

Pursuing this particular avenue of my destiny meant I was on the road endlessly between South Carolina, Washington, D.C., and Alabama. I was living in Aiken and traveling to Washington, D.C., to work as a defense contractor during the week. On weekends I would drive to Raymond's house in Alabama to conduct workshops at his Theater of the Mind. This challenging work, combined with the inordinate amount of travel involved left me living in a chronic state of exhaustion. Yet, I refused to stop. I had to fulfill the mission, and even that took on a whole new slant. I'd had an epiphany. I had the ability to help people who had been through near-death experiences, people who had no idea how to cope with life in the aftermath. Being a source of comfort and understanding for them—in addition to the bereaved I had been helping for several years—quickly became another driving force in my life. This led to my experimentation with the earliest model of the Klini, the specially created bed I saw on my first trip to Heaven. I knew from my vision that the Klini was the key to the success of the health center I was to create. I worked relentlessly with several close friends on perfecting the design. At the same time, I continued my quest to establish spiritualistic capitalism.

In Alabama, Raymond and I fashioned a modified version of the center in his home. On the weekends we welcomed an eclectic array of people interested in exploring the depth of knowledge we possessed in the recently recognized fields of near-death as well as death and dying research. Unbelievably, we attracted to us the most fascinating and brilliant individuals conceivable—all of whom fervently shared our zeal for the paranormal.

My focus was geared toward the development of the Klini. Raymond took charge of the final step of our process. He called it the psychomantium. It was constructed to simulate the energies present in the ancient Greek oracles, which Raymond called the Necromantium. The psychomantium was an isolation chamber used exclusively for monitored and sanctified communication with the dead, souls inhabiting regions of the world beyond.

In this capacity, we worked with hundreds of people who came to us from all over the world. They ranged from famous (or infamous) celebrities to average working-class stiffs, along with many college students. Regardless of age, class, or spiritual orientation, our results were phenomenal. The subjects' fear of death was usually drastically diminished, if not totally eliminated. Consequently, they no longer continued to view life as a struggle. In fact, in two short days, for the vast majority of our participants, life was magically transformed into a wondrous journey filled to the brim with endless and sparkling possibilities. Who could ask for more? It was a remarkable experience for us all.

During this time I met Paul Perry with whom I wrote my two best sellers—*Saved by the Light* and *At Peace in the Light*. The Fox Network produced a movie of the week about me. Over fourteen million people tuned in to watch the adventures of Dannion the first time it aired, and an additional fifteen million viewers tuned in ten months later. I found it mindboggling that twenty-nine million people had actually been interested in my story.

As you can imagine, my life rapidly shifted into overdrive. I was even invited to be a guest on *Oprah*—that's when I knew I'd arrived! The demands on my time quadrupled, and I was now traveling three hundred days a year, flying from one coast to the other taking off for destinations halfway around the world. My philosophy was: if they want to see me, I want to be seen by them! It was intoxicating. My newfound celebrity status caused me to lose my balance.

At the same time, I had a passion for our nation's veterans; I cared about our military personnel from WWI, WWII, Korea, and Vietnam as well as our young men and women who were defending freedom in Kuwait and Iraq. I'd served my country in the Marine Corps. That tour of duty left me deeply devoted to all my brothers and sisters in uniform. As often as possible, I would include a visit to VA hospitals and nursing homes in the cities where I was scheduled to lecture or do a book signing. I'd been a hospice and nursing home volunteer for over twenty years, and it was probably my greatest passion. I'd been there. I knew exactly what it was like to lie in bed all day just staring up at the ceiling and waiting (or hoping) to die. In the very first year after *Saved by the Light* was published, I recruited over thirteen thousand people to volunteer their time to hospice care. That was as exciting for me as selling a million books.

Still, the book tour proved to be grueling. On a daily basis, I was face-to-face with overwhelming grief and inconsolable sorrow. I had no concept of how unprepared I was to handle the barrage of forlorn and anguished people at each stop, waiting in line to meet me. People would ask me to bless their children, heal their illnesses, and take away their sins. Misery had become my new career.

As things finally started to wind down and the initial Brinkley-mania ebbed, I got a chance to catch my breath. Despite the fact that I was in a state of exhaustion, I'd kept pushing myself to meet all of my obligations, and then do a little bit more. I had this nagging feeling that my time was limited, so I had to live my life in a hurry. Inwardly, though, I was torn over living out my time without one special person to love.

My earlier experience after being hit by lightning had caused me to lose faith in love, so I armored my heart, swearing never again to fall victim to its pain and unpredictability. What's more, I was a man with a mission now. I had something to do.

So while I was off performing my usual heroics, the true powers of the universe were conspiring to bring me into a totally new state of consciousness—the realm of being married with children.

My last speaking engagement of the year was at a spiritual expo in Las Vegas. I was looking forward to it because the year before, I'd been interviewed by a journalist who lived there. When *Saved by the Light* hit the *New York Times* best seller list, I couldn't have been happier. I had done hundreds of interviews, but the one I remember the most was done by a woman named Kathryn in Las Vegas. A year later, the same journalist called when *At Peace in the Light* was released. I was thrilled at the prospect of speaking with her again. What began as a great thirty-minute interview evolved into a delightful three-hour conversation. Before I said good-bye I asked if she could possibly make it to the spiritual expo in Las Vegas.

She told me that, if she could, she would accompany her publisher to the airport the day I was arriving and we could meet there. When I arrived, I spied a limo driver holding a sign with my name on it, but no Kathryn. My heart sank. Handing the driver my carry-on bag, we headed toward baggage claim.

From somewhere in the crowd behind me, I heard my name called in a feminine voice. Turning to find the source of it, I saw a woman of medium height with dark hair waving at me. She introduced herself as the newspaper's publisher and reminded me that we had met the year before. Again, disappointment flushed over me, for there was no sign of Kathryn. Explaining that I had to retrieve my luggage, I started to walk away.

She grabbed my arm and pointed at the top of the escalator. "There's Kathryn," she said. "She came with me today to meet you." I looked up at that escalator. It was like being struck by lightning all over again. Kathryn stepped off the moving stairs, and I experienced love at first sight in every way the romance novels have ever described it. My knees buckled. My heart leaped into my throat. My mouth went dry. I felt dizzy. Kathryn walked up to me offering a red rose. In exchange, I silently gave her my heart.

From that moment, I knew life was never going to be the same again. Falling in love took only an instant. However, it would take years for our relationship to work through the web of intrigue I had woven in order to secure the success of my personal and spiritual missions.

Adding to existing complications, Kathryn was a package deal. She came complete with six wondrous children and their father, Jim (I call him the husband-in-law). Nevertheless, I was in love with all of them, and I was determined to make these relationships work.

This love was the greatest, most perfect thing that had ever happened to me; yet, in the pit of my stomach, I kept experiencing a sinking feeling, an unrelenting sense of impending doom. I knew something horrid was about to happen. I was grateful for all of the wonderful things that had come into my life and all the ways I'd been able to touch other people's lives. I'd honestly done the best I could to be the kind of person I thought the Beings of Light wanted me to be. As Albert Einstein said, "Only a life lived for others is a life worthwhile." Yet, I definitely sensed that the Beings now expected something more of me. My intuition told me that although I had been given the incredible gift of Kathryn and the gang, I was about to be put to the test once again.

SEVEN

Another Life Bites the Dust

In order to realize your destiny,
you must be willing to release your history.
—KARL SCHMIDT

It was August 1997. As I recall, it was an extremely warm and humid summer day. Having wrapped up my business on the West Coast, I walked to the waiting taxi and felt filled with a deep sense of satisfaction. Fortunately for me, my final meetings in Los Angeles went exactly as I'd planned. Twenty-two years of hard work had paved the way to this most important juncture in my life. At the veterans hospital in West Los Angeles, I signed the memorandum of understanding necessary to create the Twilight Brigade/Compassion in Action. The formation of this nonprofit, end-of-life volunteer organization had been a long and arduous task that finally came to completion with the help of a small group of people. Through the birth of this heart-child, I had achieved another mandate of the heavenly mission placed upon me by the Thirteen Beings of Light: spiritualistic capitalism, which was to be the foundation for the centers I had been told to build.

My flight to Atlanta was leaving LAX around lunchtime. Once aboard the plane, I settled into my assigned seat in anticipation of a few quiet moments to savor my recent victories. But my attention was instantly diverted by the demands of a horrible headache. It had been gradually developing over the course of the day, but I had successfully ignored it—until now.

As the aircraft ascended above thirty-five thousand feet, I became painfully aware of the fact that the increased altitude was magnifying the severity of my headache. Without warning, I felt something burst in the right side of my head. Immediately thereafter, I felt a release of pressure within my brain, accompanied by an overall sense of physical weakness. The incredible weakness was what I found most alarming. At the same time, I literally saw stars through several bright flashes of light inside my head. Then, just as suddenly, the headache disappeared—only to return ten minutes later, with a vicious vengeance. For the next three and a half hours, I endured pain beyond belief. To slightly move my head, or even slowly bat my eyelids, created a throbbing agony I was sure would cause my head to explode.

I arrived at my home in Aiken, South Carolina, after midnight. To this day, I possess no memory of how I got there. And I certainly cannot fathom how I survived through the night. Miraculously, morning did arrive. I remember pushing off the daybed in an attempt to stand as my father walked through my front door. In that instant another eruption burst in my brain. I dropped to the floor as my dad watched. When I fell, he couldn't help but see the horrific bruising from the internal bleeding that had begun in my chest and arms. As I fought desperately for breath, every feeble attempt to form words revealed an escalating slur in my speech. I honestly believed this could be my last day on Earth. I knew without a doubt that either I had a blood clot or an aneurism was rupturing in my brain. The panic I witnessed in my father's eyes filled my heart with the wrenching pain of regret. How I wished he did not have to see me like this again. This scene had become far too familiar.

My father has always been my hero, and in that moment, his great sadness was more than I was able to bear, so I closed my eyes to hide from it. Inwardly scrambling, I sought to find a modicum of comfort in my recent spiritual triumphs. Nothing would stop the momentum of what I had put into place with the creation of the Twilight Brigade. American veterans of war now had a powerful support system of appreciative volunteers devoted to being at the bedside of our nation's heroes. Indeed, if this was to be my final exit, I would make my transition with pride in the knowledge that grace was firmly anchored in place.

Doing all he could not to collapse emotionally, my dad just wanted me to get to the hospital. After he helped me from the floor and back onto the daybed, I asked him to just let me lie there long enough for the dizziness to pass. At this point I was having increasing difficulty remaining conscious. Although the pain in my head was blurring my vision, I managed to dial the telephone number of a friend who could drive me to Charleston. I knew my dad was simply not up to it. The two hours it took to drive to the hospital in Charleston seemed like two years. We phoned the hospital so the doctors and emergency room staff would be aware of our arrival.

I can recall only a few bits and pieces of what transpired over the following two or three days. And even those tidbits are not clear, for the endless pain left me dazed as I drifted in a medicated fog from one state of consciousness to another. I do remember being told that the tests confirmed I had a small brain aneurysm as well as three subdural hematomas in my right temporal lobe. Basically the aneurysm was inoperable, and the subdural hematomas posed an imminent threat to my life. With every heartbeat, a copious amount of blood was being pumped into my cranial cavity. This created an unbearable pressure pushing my brain against the left side of my skull, and continued pressure like that would surely kill me. Although the medications were, for the most part, ineffective, they remained the only small token of comfort the doctors could grant me. This torture continued for days on end. A surgical procedure to relieve the pressure and alleviate the pain was possible, but not while my blood was so thin.

Since my open-heart surgery, I had been on a high daily dosage of Coumadin to thin my blood. Because I had consumed up to 30 mg per day (the average dose is 2–5 mg per day), my blood was so thin now that I would most assuredly bleed to death in the operating room. I had no choice but to stop the blood thinners and wait for my blood to thicken sufficiently.

Desperate for a distraction after too many days of confinement, I made up my mind to bust out for a while and convinced a friend to act as my accomplice. After helping me escape the hospital, my friend drove me directly to the beach. Sitting in quiet solitude before the limitless, azure expanse of the Atlantic Ocean did me a wonder of good that day. My spirit took flight far from all that worried me. I felt the freedom to soar above the Earth and really breathe again. From this exalted vantage point, I spent the next few hours in an honest reflection of my life. Looking back over my entire life, in the safety of the stillness, my mind basked in the light of so many events that made me proud. Yet, there were many other things I could have done differently and so much better. In the glaring spotlight of this brutally honest self-assessment, my life made me smile, despite the sprinkling of sadness and regret. Actually, I had come a long way since my days as a thoughtless jackass. Well, maybe I was still a bit of a jackass, just not as often. All in all, as I sifted through my earthly existence, I achieved a serene sense of spiritual completeness. If my time was indeed coming to an end, thanks to this day of contemplation, I would go peacefully. But if my instincts were right, things were about to go from very bad to much worse.

By my second week in the hospital, my sense of time no longer existed. Nothing made the pain go away, and it was stealing my strength. It's amazing how unrelenting pain can ruthlessly wear down the spirit and short-circuit all mental processes. Determined to battle till the end, I fought to remain conscious and fully aware of changes in my body. To achieve this, I listened endlessly to healing tones or white noise through a headset in order to drown out the cacophony of the health industry operating just outside my door. Inside my mind, I existed in a meditative world infused with the serene presence of spiritual peace. This is a world I'd been introduced to during my first near-death experience. Since that time, I had made a point to revisit it as often as possible.

By deliberately placing myself in an altered state of consciousness, I could enter a dimension to escape the pain for short intervals of time. During these out-of-body journeys, I learned to recharge my physical self with the vitality found in my etheric body, where pain and discomfort are not a reality. I accomplished this by focusing on a single sound and making a tone in my throat. Then I would visualize a color. For me, indigo or violet worked best due to their remarkably high spiritual frequency. I'd hold the one color in my mind's eye until it saturated my being and, flowing outward, it would eventually fill the entire room with its healing frequency. When this exercise is practiced on a regular basis, it takes less and less time to arrive at that space between the worlds.

This out-of-body technique was the only thing that made it possible for me to endure the first few rounds of this heavyweight championship fight for my life—a fight the doctors told my father I only had an 8 percent chance of surviving and really no real chance of winning. There was no doubt in anyone's mind: I was going to die. At this point, radio host Art Bell and his beloved wife, Ramona, had traveled to Charleston, all the way from Nevada, to be at my bedside. I can never express how grateful I was to have them there for me.

Finally, the doctors concurred that the only way I could possibly make it through the surgery would be for them to cut open my chest at the same time they removed a sizable portion of my skull. This way they could monitor my heart for any clots that might find their way to the artificial valve I'd received in 1989. I didn't like the idea of this process at all! Could you imagine the thought of facing brain surgery and open-heart surgery at the same time? No wonder they only gave me an 8 percent chance of surviving.

Then I was granted an eleventh-hour reprieve—the double surgery was scratched. In a stroke of genius, one of my doctors received what I believe was a divine inspiration. He proposed an alternative procedure to drain the pooling blood in my brain by drilling three holes in the side of my head. The options were clear: relieve the stress to the tissue or contend with the possibility of irreversible brain damage. After a brief discussion with my family concerning the risks involved, we made the decision for the doctors to move forward with the procedure. It took just over two and a half hours to drill the holes and insert the drainage tubes. The surgery went well—I was still alive and the blood was draining. However, things got a bit scary after that.

For the next forty-one hours, my lifeless form laid in the recovery room while I struggled to regain consciousness. My spiritual self was wandering the netherworld in search of understanding, which I will describe in the next chapter. During this time, the nurses worked to awaken me, and the doctors grew gravely concerned. When I did come around, a very relieved nurse welcomed me back to life and soon moved my bed to a private room.

Within the hour, I had a massive grand mal seizure. In reaction to the terrifying hallucinations that commonly accompany this type of seizure, I tried to escape the hospital. In my efforts to free myself, first I pulled out one of the drainage tubes from the front of my head. Next, I violently ripped the intravenous tubes out of both arms. Throwing the bags of antibiotics and saline solution on the floor, I made my way to the door. Two strong young orderlies came charging in to stop me as I attempted to leave. It took the best efforts of them both to restrain me. The last thing I recall is being tied to my bed by the two orderlies while the nurse tried to replace the drainage tube. Then everything faded to black. I slipped into a coma-like state for several hours and awoke to find myself still strapped down and in a neurocardiac observation ward. The doctors confided in me much later that I had been moved to that ward because grand mal seizures often precipitate a heart attack, a stroke, or another grand mal seizure, which is often fatal.

Through divine grace and the loving prayers of others, I was saved from an even worse fate. Slowly, I began to regain my strength, and further surgery was deemed unnecessary. After close to thirty days in the hospital, I was finally released to go home. But recovery from brain surgery is a long and difficult process. The doctors told me I would probably experience headaches for a couple years following the operation, and they were right. When those headaches would hit, my life would come to a screeching halt. The searing pain would be so debilitating that I often experienced extreme dizziness and loss of balance.

The good news is, as the years pass, I feel a little better with each new day. Occasionally, a sharp pain or headache will remind me to stay on course. But by and large, the condition of my health is good, although still challenging. I still have to deal with various physical problems, considering everything that has happened to my body. Yet, even in light of all my personal ordeals, the wonders of life never cease to enchant me. In fact, I would like to share some intriguing details for you to ponder: I was struck by lightning during a severe thunderstorm at 7:05 p.m. on Wednesday, September 17, 1975. Twenty-two years later, during a severe thunderstorm, at 7:05 a.m. on Wednesday, September 17, 1997,

I was scheduled for brain surgery. Wow, what an ingenious plan Spirit formulated to get my attention! What do you think the universe was trying to tell me with this awesome display of cosmic synchronicity? Of course, maybe this could just be chalked up to a mere coincidence. I do believe in coincidences; I've just never seen one.

EIGHT

Varying Degrees of Heaven

Life is pleasant.
Death is peaceful.
It's the transition that's troublesome.

—Isaac Asimov

During the operation to drill holes in my skull, I returned to the heavenly realm and discovered some things I had not noticed before. I debated over the years about whether to disclose the details of my third near-death experience, as I could find no mental framework appropriate for conveying the disturbing details that demanded to be part of the story. What my soul experienced while my body underwent brain surgery was dreadfully discombobulating.

In thousands of lectures, I had assured the listeners that there was no such thing as Hell. I rationalized this quite aptly by saying to my audiences, "If I didn't go there, with all the trouble I've stirred up in my life, then surely none of you are going!" I genuinely believed this statement to be a verifiable fact. In my previous visits to Heaven, I had personally witnessed only states of love and levels of bliss. And after I'd had the time to properly assimilate all the details of my experience, I was happy to report my findings to the world. I'd gained this perspective with the greatest of ease. Never, in my wildest imaginings, had I dared to conceive there might be more to the Hereafter than what I had been shown in my first two near-death experiences.

Bear in mind that in both my previous visits to the other side, I remained only vaguely aware of the existence of different levels of Heaven. During my visit to the Crystal Cities, I did look down, quickly observing others who appeared to be lost. It was obvious they were vibrating at a lower rate. However, I paid little attention because just looking at them caused my vibration to slow to a rate that matched theirs, which was very uncomfortable for me. Moreover, I wanted to concentrate my attention only on the level I occupied. Whatever was above, below, and on either side of me held no real interest. Believe me, there is nothing odd about that. My general response to life has always been, *It's all about me!* So in my mind, what was happening to any other soul on the other side was absolutely none of my business. I just wanted to know where the Beings were taking me and what we were going to do when we got there.

Remember, the Bible tells us there are seven Heavens. As long as I was already there, I was ready to see them all! In my first two experiences, I was taken through glowing portals of crystal and into the Crystal Cities. Then, right before my spiritual eyes, thirteen Beings of Light appeared behind a glistening podium. Immediately, I sensed they were thirteen genderless Beings of great power. Each of these Beings was the celestial embodiment of what I would call a human virtue, such as heroism or wisdom. In their presence, I did not just learn, I became knowledge. One by one, the thirteen Beings of Light infused me with visions that would shake the world in the not too distant future. After my recovery permitted it, I wrote down the visions I'd witnessed on that unforgettable day in our Father's house. The Bible tells us, "In my Father's house are many mansions" (John 14:2). I now know without question, this is truth. But I'm still having a little trouble with the Biblical bit about the streets of Heaven being paved in gold. Nobody ever showed me any of that, and I was really looking forward to it!

Anyway, what I want to share with you is how my third near-death experience compared to the first two. In many ways it was similar because I experienced a lot of the same feelings and sensations. But there were also some strikingly dissimilar aspects waiting to greet me on my third visit. I believe my sojourn to the great beyond during brain surgery in 1997 was actually a culmination of the two previous experiences. After much thought, I've concluded that the first death experience in 1975 (I was dead, take it from me, there was nothing *near* about it) was intended to acclimate me to the process by which we all make our transition from this world to the next. This is something I want you to grasp: Like most of us, until I was twenty-five years old, I was convinced that my world existed within a reality that was mental and physical in structure. However, upon my death, I gained access to a multidimensional reality. Within the context of these realms, I reconnected with what would later become my well-documented empathic/psychic skills. I believe we all innately possess such skills, along with the potential to hone them.

The content of the second experience brought into my consciousness an even more keen perception of the multi-universe. I was educated regarding specific future advancements in science (nanotechnology) and medicine (photon therapies). The Beings showed me how innovations in these two areas would be instrumental in the process of stress reduction. I recognized that I would incorporate them in the centers I would be sent back to build. With stress being the number one cause of disease among humans, the Beings compelled me to initiate the centers as a form of spiritual activism, instituted to help us harness the healing properties of serenity.

Continuing my education in the third experience, the divine scholars chose to amplify my knowledge. This time they took on the task of explaining to me the awesome hierarchal configuration of the celestial realms. I'll never forget the wave of relief and joyfulness that washed over me when the lesson came to an end. For as a result of what I was shown, I can tell you unequivocally there is no Hell, only varying degrees of Heaven! From states of ecstasy to sobering states of self-contemplation, the heavenly realm functions in every way to bring comfort to the soul.

In all three experiences, I lifted out of my body and journeyed through the blue-gray place where I believed myself to be the only one present. I dwelled completely in this nebulous space, yet I could still clearly sense the world I had just left behind. The tranquility surrounding me was bathed in the soothing sound of silence—for a while.

Before long, I became aware of the sound of chimes coming from far away. Seven chimes, to be exact. The melodic echoing created within me an exhilarating sensation of lightness and expansion. As the chiming continued I vibrated in harmony, becoming less and less dense. Now as light as air, I felt myself lifting through higher dimensional octaves. In retrospect, I now recognize that the chimes set the tone for the elevation of my consciousness through these different spheres. Empowered by the vibration of sound waves, the chimes literally opened the spheres, enabling me to shift effortlessly through their successive levels. Prepossessing this awareness from the first near-death experience helped me during the second experience to more fully appreciate the beautiful precision deftly at work in Heaven. I enjoyed being an eyewitness to the graciousness of the system by which we take leave of the present life. It is a system superbly engineered to assist us in opening, once again, to the truth of our divine identity. This same system infused me with the stunning revelation that we live in different dimensions, and that many realities exist beyond this one.

At the same time, I saw so clearly that our mortal, human presence within this earthly plane is in no way diminished by the presence of those other realities. In fact, it was indelibly impressed upon me just how vitally important the work we do on Earth is in the greater scheme of things. In earnest, we need to take to heart this one fact that the evolution of every one of us, along with humankind as a whole, has an unmistakable and mighty impact on the overall development of the entire universe. Each of us is intricately woven into the fabric of God's ultimate plan of perfection. Collectively and individually, we are indispensable in the pursuit of that perfection. Although we are all unique in our identities and how we manifest our life force, we remain an inseparable component of the one consciousness—our universal unified field. Like the individual notes in an exquisite melody, each of us is essential to the flow and the harmony, to the wholeness and the beauty of the universal rhapsody that breathes life into the core of our souls. Even more astonishing for me was the realization that, as multidimensional beings, aspects of us (in our multilayered totality) are doing this powerful work in several places at once. I don't know about you, but I thought that was some pretty wild and cosmic information to grip!

I always thought one life held quite enough spiritual intrigue for me to handle. Then to discover we coexist, with different aspects of ourselves, in multiple lives on multiple levels of consciousness was a huge leap for me. Nevertheless, I knew it was true. And it helped me to understand why we need so much sleep at night. Think about it—with all the work we're doing on so many levels, no wonder we feel tired all the time! This bounty of universal truth was the treasure of knowledge I acquired from my visits beyond the veil.

Everything in the third near-death experience remained pretty consistent with the first two—to a point. The places I traveled to and the new sights I witnessed in the third jaunt to the afterlife created an entirely new paradigm of what I had believed to be the nature and structure of Heaven. I was unexpectedly forced to accept a more comprehensive view of the Hereafter, like it or not. Consequently, I felt ambushed.

Perhaps *betrayed* is an even more appropriate word. In no way was I adequately prepared for what I witnessed. In all honesty, I simply wanted to forget about it. So, like many of us, I used work as an escape. For me, it was hospice work. I spent far too many hours at others' bedsides. I rationalized that by vigorously striving in the direction of my divinely appointed mission, I could bypass the need to divulge this new information to the public. Yet for those very close to transition, I did everything I could to help them avoid being trapped in the places I had seen.

However, much to my chagrin, I discovered the universe has a way of being irritatingly persistent. That still, small voice within adamantly refused to give me a break. At regular intervals (about every two months), I was aggressively prompted by my guides to share this information.

While we're on the subject, allow me to take this opportunity to emphasize that we all are surrounded by the loving protection of entities who are sworn to watch over us, day and night. Often referred to as our guides and guardian angels, these souls may be our departed loved ones—friends or family members who've made their transition—or they can be highly evolved entities with no earthly ties to us.

Well, some of my guides I like and some I don't. I especially hold little to no affection for the ones who like to nag. They kept nagging anyway, and I failed, one pitiful attempt after another, at formulating a balanced way of teaching this newly revealed truth. How could I tell this story without adding to the stress and worry in people's lives, particularly in the lives of those who were already dealing with the end of life or the loss of a loved one?

Then I was blessed with an insight: These were the people who would benefit most from this information. Instead of creating more stress in their lives, what Spirit was asking me to offer was a spiritual truth as a means to circumvent undue worry. I had been given the arduous task of breaking stunning news to the world. Initially, it seemed harsh, but through the firm guidance of my Angelic Helpers, I came to the realization that it would eventually help millions of people improve the quality of their lives, here and in the Hereafter.

I ask you: does Spirit have a sense of humor or what? Think about it. Here I was, an authentic hellion, a rarified product of religious rebellion. I was born and bred in the Bible Belt, in the heart of the Deep South. Why in the world had I been selected spokesperson for the other side of the veil? Why was I chosen to share with all humankind the news flash of the new millennium?

I wrestled with this question over and over again. Can I tell you something just between good friends? I have never been able to come up with a good enough answer. The one thing I know for sure is this: one day I never believed in any of this stuff, the next day I was dead, and the day after that I became everything I never believed. I also know something else I must share with you: we dwell in a fair and just universe. I have been shown repeatedly that all things do truly work for good. Therefore, I have total faith that what I share next will edify your mind and uplift your spirit in a way that will bring true peace to your soul.

NINE

The Fertile Void

Death is more universal than life;
everyone dies but not everyone lives.

—A. SACHS

The third time I ventured heavenward, just like before, I went directly to what I refer to as the blue-gray place. The other times I was aware of it as an empty space, and I moved through it rather quickly. This time I felt as though I'd been deposited there, and I realized I was not alone. It quickly became apparent to me that this deep sanctuary was actually, in and of itself, a completely separate level of consciousness.

I had never before discerned the blue-gray place as a sovereign reality poised between this world and the next. Standing at the edge of the ominous environment, this semi-dense dimension weighed heavily on my senses. Reticently I admit I can only compare it to what Catholicism refers to as purgatory. The Holy Roman Church considers these various sublevels to be where those considered already saved must undergo purification to achieve the holiness necessary to enter the joy of Heaven. In fact, the Greeks, Assyrians, Egyptians, Babylonians, and even the Mayans all wrote of the different levels existing between the here and Hereafter.

Personally, I've come to terms with this netherworld as the place where we can dwell as long as it takes for us to recognize ourselves as immortal spiritual beings rather than physical and mental beings. It is not where we are punished for our sins; it is a region of celestial consciousness where we purge our personal misconceptions of reality, based on free will. Here we are capable of releasing ourselves from all earthly misconceptions in order to reconnect with our essential purity, our supreme nature. Safe within this fertile void, souls are watched over and supported as they go through the process of shedding human characteristics and attitudes that impede the expression of their authentic divinity.

During my encounter with this mystical region, what captured my attention first was the way the thick, slow-moving energy zapped my strength. It left me feeling weak and anxious. Then I was abruptly overtaken by the uncomfortable presence of countless souls milling nearby. They seemed caught in a vicious repetition of recycled depression, dejection, and desperation. After a period of intense observation, I was suddenly impressed with the details of their stories. These souls were reliving their last days on Earth, over and over, without end. I was infused with the knowledge that some of these lost souls had been trapped in this heartbreaking shrine of emptiness, replaying the bleak misery of their lives for what felt like hundreds of years.

One such heart-wrenching scenario I viewed was that of a soldier still wearing his tattered Confederate uniform. I tuned in to the overwhelming anguish he had endured as he clung in desperation to the side of a wooden barrel. The soldier appeared to be floating in the Atlantic Ocean, just off the coast of one of the southern states. Again, I was infused with understanding. He had been holding on to the barrel, convinced that someone would soon be coming along to rescue him. I could feel his resentment building to a breaking point as he grew weary of his watery prison. The dejected soldier refused to let go of the barrel because his pride held the righteous stance that he had been honorable in defending his principles. By virtue of that, coupled with the personal sacrifices he'd made for his country, he trusted someone would come along to save him from this undignified demise. Of course, the truth was he simply could not accept the fact that his days on Earth had come to an end. It was painfully obvious the soldier was refusing to relinquish what he considered unfinished business—and the unfinished business happened to be *his life*. It never occurred to him he was dead. And I didn't know what would have to happen for him to finally accept it.

Looking around me, I spotted many more soldiers in unique, yet similar, circumstances. They wore a variety of uniforms, though I didn't recognize many of them.

They appeared to represent all the wars in our world's history. These men were grouped together, but not necessarily based on the specific war or conflict in which they had so valiantly fought. Instead, they seemed grouped together according to the mindset they possessed toward their participation in battle or their attitude over the devastating loss of life they had witnessed. Of course, some of these brave soldiers had been killed so quickly or unexpectedly, they'd literally had no time to mentally register what happened. Others had been forced to endure horrific torture and deprivation prior to their death. Finding it impossible to accept the unthinkable inhumanity thrust upon them, these soldiers were unable to reach any tangible level of forgiveness.

Ironically, forgiveness was precisely what was needed if they were to find a state of grace. For only through grace could these lost souls discover the path waiting to lead them to the light.

Not far from the area where the soldiers were clustered, another tragic scene materialized before my eyes. It was as though this sequence of vignettes had been meticulously staged solely for my spiritual edification. Within seconds, I was surrounded by thousands of innocent men, women, and children whose lives had been snatched away ruthlessly. They were guilty of only one thing—being in the right place at the wrong time. Their lives were annihilated in senseless acts of military violence, the same kind of merciless carnage wars have perpetrated upon humanity since the advent of civilization. Can you imagine sitting down to a lovely dinner with your precious family, and the next thing you know, the house is blown to smithereens by a five-hundred-pound bomb? This brand of unconscionable hostility, in the name of freedom, conquest, religion, and colonization has happened far too many times in recent centuries. This forlorn collection of souls was still frantically searching for the reason they had been robbed of their lives. Inconsolable children with tear-stained faces wandered aimlessly, calling out for parents they desperately longed to find. Heartbroken spouses and grief-stricken parents rummaged in search of the loved ones from whom they had been so brutally separated. Their minds could not fathom what had happened to them. I was dazed by the throngs of unsuspecting family members who, as casualties of war, had been instantaneously transported from the vivacious, Technicolor realm of the physical to this thick, murky blue-gray depot of desolation. At this point, I was consumed by their panic and fear. How could this happen to the blameless? Why were these souls suffering at the hands of man-made evil? Why hadn't God intervened and stopped their agony?

Before receiving any kind of answer to my questions, another sector of the blue-gray quagmire made its presence known to me. I found myself standing in the mists of a collection of women hailing from a sundry of historical eras. In all shapes, ages, and fashions, the women appeared to be gathered under the umbrella of one common characteristic—the loss of their self-respect. I was given the mental understanding that these ladies had all led lives in which they had been victimized, abused, demoralized—although, on some level, they knew they had the option to end the cycle. A palpable torment permeated the air. Yet, in the misery of martyrdom, they believed they had found their earthly purpose. Their intimate relationships, besieged with bitterness and blame, had indefinitely delayed their divine inheritance—a hallowed place in the afterlife. Again, my heart sank into a sea of helplessness. What could I do about this overwhelming anguish? I sensed I was being pulled out of the deep silence of the blue-gray place, and I started to hear the sound of voices. Someone was calling my name. I opened my eyes to find myself in the recovery room. The brain surgery had been successful, and I was back in the earthly realm.

Yet, for many months afterwards, I could not shake off the chilling remnants of my time in the blue-gray place. My heart held a constant ache for all of the disenfranchised spirits I came across in the nearly two days I spent there. Ceaselessly I sought to understand the reason those people were held hostage to agony. All victims of a devastating absence of hope, these souls had lost sight of their self-respect, refused to express a genuine appreciation of life, or sacrificed the refuge of a meaningful spiritual foundation. Thus they had also collectively detached themselves from the security of their eternal source. I'd watched them amble aimlessly through the damp mists of the world between realms. Each was waiting for the sun to shine on them once more. I had to know why.

Then, one afternoon, while still recovering from the surgery, I was absentmindedly leafing through the Bible when my eyes fell upon Mark 10:14–15. Here Jesus says, "Suffer the little children to come unto me, and forbid them not, for of such is the kingdom of God. Verily I say unto you, Whoever shall not receive the kingdom of God as a little child he shall not enter therein." When I read this scripture, I remembered it from my Sunday school classes. In an instant, the proverbial lightbulb flicked on in my head. At last, all of that infernal Bible thumping from my childhood was starting to make some sense. In these few sentences, I found one of the greatest secrets of the light. Or more aptly put, I discerned exactly what was keeping multitudes of people far *from* the light. This short scripture gifted me with the epiphany that had eluded me for so long.

I finally got it! The trapped souls I beheld between Heaven and Earth were not being detained against their will; they were held *by and according to* their will. The agony they experienced was of their own making. To enter the Kingdom *as a child* one must possess, like a child, a heart pure and true. In other words, one must have an open, loving, cheerful disposition toward life and all the while maintain a belief that life is a gift to be generously shared and vigorously celebrated. With the comprehension of this one biblical perspective, I understood. I believe this scripture was actually a call for all of us to return to our divine innocence, our ability to view life only through the eyes of love. For the trapped ones, only their inner light of hope, faith, and love could act to dispel the clouds casting the dark shadows over their self-imposed exile. God had not abandoned them. Quite the contrary, these souls had abandoned faith in themselves and in the goodness of life. A renewed faith and commitment to joy was all they needed to free themselves from the bondage of this hellacious separation from the spiritual kingdom. Without proper spiritual grounding, what had happened to them could happen to any of us.

Although this striking insight brought me a sense of comfort, I continued to be plagued by sleepless nights spent mulling over the blue-gray place. *How* did it exist? What made such a place possible in the spiritual hierarchy? I couldn't stop thinking about it. The more I thought about it, the more outraged I became at the active role our governments, institutions, and religions have conspired to play in the spiritual bankruptcy of America. Because of their failure to inspire a living faith in a fair and just system that protects and guides us as we leave this world, we have all suffered immensely. As a nation, we have unwittingly allowed this soulless agenda, contrived and perpetrated against us, to successfully disconnect us from eternal destiny. As an unfortunate result, the vast majority of us spend our entire lives trying to avoid the inevitable. Or we hope, with our last breath, that somehow death will pass us by. In Heaven, I was shown the essential part death actually plays. It is an invaluable facet of the eternal, and this truth needs to be inculcated from childhood as a core belief, taught as part of our education just like the ABCs and multiplication tables. The way I see it, death was conceived by God to be the ultimate life coach.

As such, death must become one of life's mandated conversations. Did you know that 72 percent of families living in the United States never discuss death until they lose a family member? This is an atrocity. It is denial at its zenith. We must initiate this conversation. Perhaps if we do, we might be able to help free those souls already trapped in the blue-gray place. At the very least, we will be able to help decrease the number of souls who pass time there.

As frightening as the last thought I need to share with you is, I feel compelled to tell it. After my extended stay in the blue-gray place, I realized that if the number of souls there continues to grow at this rate, within the next couple of decades we will have a very serious afterlife disaster on our hands. You see, it is already filled with millions of souls. A few million more will create a critical clog. At that point, one of two things can happen: either the souls already in the blue-gray place will seep back into this world, bringing with them tremendous misery and fright, or the souls leaving this world will be unable to get through the tunnel and into the light. Either way you look at it, this is a prospect beyond terrifying! To be completely honest, I think the backlog from the netherworld has already started to overflow into our reality. This is validated by the recent reports of a 700-percent increase in exorcisms performed by Catholic priests over the past decade.

Teenage suicide is on the rise as well. The psychic attachment of dark souls could be one possible explanation for this depressed and self-destructive trend in adolescent behavior. They are the most vulnerable and susceptible to spiritual attachments and possessions due to their lack of identity and self-confidence, which often leads to experimentation with drugs and alcohol.

For discarnates wandering in despair between the worlds, life is a joyless struggle. This eclectic group of souls is spiritually united by a shared dismal outlook on life. They display ongoing cynicism, apathy, a sense of betrayal, as well as a refusal to take responsibility for their own happiness. Somewhere in the course of their physical presence on Earth, they made a decision to surrender their passion for living and to reject the search for a spiritual identity that would make them complete. In one way or another, each of them had stopped living long before they died. Paradoxically, even in death they refuse the gift of life. There is no doubt in my mind that these trapped souls are responsible for the ghostly aberrations and hauntings that have been written about through the centuries. Chained in spirit to the destructiveness of their adopted negativity, they interminably replicate their past life patterns en route to nowhere. My heart breaks for them still. If only they could see, with a few faithful, optimistic thoughts focused in the right direction, they could trade in this state of nothingness for true peace in the light. If there is just one lesson from Heaven you could commit to memory, let it be this: our thoughts create our attitudes, and together they create the quality of life we experience—both here and in the Hereafter.

Luckily, for me, my visit to the blue-gray place was short, but it was long enough for me to get the point. I'd been detained there for a reason. Searching my soul, it didn't take me long to realize that I too was becoming trapped in my own cynical and apathetic ways. I understood that I was being given a sneak preview of my own destiny, should I choose not to change. For far too long, I'd rationalized my attitude. With all I had seen in my life, I figured I'd earned every right to be skeptical, suspicious, and brash. But the truth is, I was wrong.

If anyone was familiar with the delight of roaming the universe and the thrill of rummaging through Heaven, I was. There was no way I could be exempt from the responsibility of owning my spiritual joy or of living with zest, curiosity, and openheartedness. The recurring patterns of life can undoubtedly be hypnotizing, and as a result, I had fallen into the working man's trance wherein I stopped appreciating the obvious. I had allowed the temporal world to taint my mind and harden my heart while my eternal reality conveniently slipped from my mind.

The third near-death experience was by far the most sobering yet enlightening one of all. I realized, for the first time, that we do not necessarily leave this side of life and end up in the glory of Heaven, as angels of infinite light. In my previous experiences, I'd been ushered straight to the presence of divinity. I was shown the future and the part I was to play in it before they sent me back. But on my third visit to Heaven, I learned that when we leave this side of life for good, we pick up life on the other side with pretty much the same issues and attitudes to work through. Our quest for spiritual perfection is an ongoing process, both here and in the Hereafter. What we leave unfinished here, we will finish there. So, it really does serve us to live in a way that honors the gift of life we have been given.

Today, not a single moment passes when I am not aware of the preciousness of my life. I received a personal and much-needed wake-up call in seeing exactly where I would find myself if I refused to stay anchored in the light. Also, once I grasped the disparaging reality of being trapped between worlds, I was given the opportunity to come back and tell you everything I'd been shown. From the bottom of my heart, I pray that you learn from my experience and eagerly create, for yourself, the kind of loving reality that fosters the magnificence and divinity of the gift we call life.

PART 2

The Fourfold Path to Power

TEN

And Now, the Rest of the Story

I've never seen a monument erected to a pessimist.

—Paul Harvey

Just in case I have almost scared you to death with the last chapter, allow me to put some life back into all of this. My major focus for the rest of this book will be to teach you how not to be afraid of what I've already shared with you. In fact, I'll teach you how to capitalize on it. If there is anything I know for a fact, it is that really horrible events can be turned into something truly wonderful. These events can serve as catalysts for major changes in our life perspective and as teaching tools for helping others.

For me, it took two lightning strikes, open-heart surgery, and brain surgery to see that I had more to contribute to the world than just a bad attitude. But then, Mother always said I was a little special. I guess that meant she knew, before I did, that the only way I was going to learn my lessons in life was the hard way. To be completely honest, I don't think I really started to learn any of the lessons presenting themselves to me until after the surgery in 1997. Before that, every time I was slapped in the face with another catastrophic life event, instead of getting smarter, I just got mad. *Mad* isn't even the word for it. I became enraged. And the more enraged I became, the more I attracted challenges that were more difficult and menacing than the time before. Without a doubt, I was living proof of the principle *What you resist persists.* I resisted personal and spiritual growth, and the universe persisted in its efforts to teach me in spite of myself.

I would like you to benefit from all I have learned without having to go through what my stubbornness forced me to. As I've already shared, the surgery in 1997 brought me to my knees, forcing me to revamp what I always called my automated systems. When I finally concluded it was to my advantage to view life from a more spontaneous and spiritual point of view, the most wondrous changes began to take place right before my eyes. For the first time, I permitted myself to set sail into the waters of the mystical. I enthusiastically embraced my life, not for my purposes, but rather in compliance with the will of Heaven.

Prior to this, the near-death experiences had served to evoke resentment within me. This resentment always resided just beneath the surface of my gregarious façade. In truth, I felt as though the entire lightning experience had ultimately robbed me of my identity. I wanted to go back to being the person I was before. I wanted to do exactly what I wanted, when I wanted to do it. I mourned for my lost sense of autonomy. As a spiritual by-product of bearing witness to the afterlife, I was expected, if not strong-armed, into being someone I never desired to be. When my public notoriety soared after writing the first book, my frustration intensified. The pressure was on to perform as this other person—a person foreign to my natural self. I intuited a subliminal message from the people around me, an expectation for me to be saintly or as angelic as my hosts in Heaven. In many ways, I watched my basic personality eradicated in favor of what others needed from me. Needless to say, I didn't like it.

Equally disconcerting were my ever-increasing psychic abilities. Often, I thought these skills were a curse disguised as a blessing. A dear friend and renowned parapsychologist, Dr. William Roll, later explained to me that after I tore down the veil between the worlds, it was impossible for me to tell where I ended and other people began. Imagine what it was like living in a constant state of no boundaries. It was a wretched condition of psychological bewilderment that caused me unbelievable stress. I had no choice; I just had to deal with it. The Beings of Light had set me on a path that exacted great courage and most assuredly promised to lead me to new pinnacles of compassion, principle, and integrity. However, at the time, those were not virtues that particularly interested me. Now, as I stand on the other side of myself, I can detect all of the many ways my inner self maneuvered my arrival to this fated level of spiritual maturity. So if you happen to be facing the same dilemma, take my advice: get on with it, and the sooner, the better. Trust me, no matter where you run, every path still leads back to *you!*

I grew strong spiritually and physically in direct proportion to my ability to set aside my feelings of victimization over what I had believed to be my undeserved share of life's pain and suffering. Exchanging feelings of self-pity for an authentic appreciation for what life was offering accelerated my spiritual growth. *A Course in Miracles* tells us a miracle is just a shift in our perception. Making the conscious choice to respond to life instead of reacting to it produced an inner balance I had never known. More and more, I made the choice to view situations through the lens of love, as opposed to the cold-as-steel outlook of my former cynicism.

Do not think for a moment that this process of transformation was easy on me. It wasn't. I took one step forward and about five steps back for quite some time. After three near-death experiences, the most important thing in my life was my commitment, my sincere desire to grow. I was determined to see it through. Set in my ways and mean as hell, I had always demanded that life meet me on my terms. Going with the flow and proceeding as the path unfolded were disturbing and complicated concepts for me to adopt. Without being in total control of every detail of my life, I was miserable, feeling way too vulnerable and exposed. Yet, regardless of the raging discomfort, with each new day, I dedicated myself to the process of self-change.

Once I decided to become an active participant in overhauling my own life, I came to quite a few thrilling realizations. First, I discerned that the inner workings of life transitions are spiritually elegant, simple in form and perfectly structured. I could see clearly how the process of personal transformation is predicated upon our unique spiritual strengths and weaknesses. It is not an external system placed in opposition to our personal will. Quite the contrary, it is a natural sequence, internally designed to support us in accordance with the universal laws of personal growth. Conscious cooperation with spiritual transformation is the best way I've found to proactively live a rewarding life in a manner that honors the eternal soul.

In seeking to comply with the means of my personal maturation method, I have come to the understanding that certain perspectives help us to benefit greatly from every life event. The adoption and incorporation of these outlooks are now so vitally important that they are perhaps even tantamount to our planetary survival. It is my conviction, based on the scenes from the Boxes of Knowledge, that by 2012 humanity will experience unprecedented mental and spiritual transformations, coinciding precisely with the Earth's passage through great physical upheaval. I sincerely believe a dramatic shift in consciousness will occur within this designated time frame in order to set the stage for what is to come in the years directly following.

In this new state of being, we will embrace an awareness of the companion spheres of life all around us. And within that awareness, we will possess the knowledge of our simultaneous existence within these multidimensional realities. I swear this is not as weird or fantastic as it might sound. In fact, with the advent of quantum mechanics and physics along with the evolution of the string theory and our comprehension of the chaos theory, what I am describing has been a part of our everyday reality since the last quarter of the twentieth century (when we successfully landed men on the moon and invented the PC, cell phone, and DVD). Personally, I know these parallel dimensions to be real because I have been conscious of existing within them for many years now. In the early days, while I was recovering from lightning, orienting myself to more than one dimension was extremely tedious and tiring. Now, it is just part of my everyday reality. I can become as involved or stay as detached from these realities as I like. It totally depends on what kind of mood I'm in on any given day!

However, before humanity as a whole is elevated to such a comprehensive multidimensional awareness, it is imperative that certain concepts be understood. There is no doubt in my mind that collectively we have the power to change the world for the better. We simply have to collectively agree on the changes we want to see. Mahatma Ghandi said, "We must become the change we wish to see in the world." So, if we were all to become the changes we collectively desire, imagine what the world would be like. I think about it all the time: a world without war, a safe place to raise our children, a world devoid of all environmental pollution, a place where equal opportunities are available to everyone and freedom overlooks no one. How do we achieve this? Do you even believe it's possible to create such a world? I believe it is. The visions showed me that it is.

Remember, the Beings of Light said that history is not carved in stone; we have the power to change the future. Yet there are more than a few changes we need to make in order to do so. A multitude of concerned and effective organizations are standing by to enlighten us as to the present condition of every ecological issue, from our skies to our oceans, as well as all the endangered life forms inhabiting them. I urge you to contact a philanthropic society that speaks to you and to volunteer a portion of your time weekly in order to empower yourself and a cause you believe in. Undoubtedly, this will be the most valuable form of spiritual activism over the next few years. In addition, go green and learn everything you can about sustainable living.

My personal commitment is to all initiatives necessary to ignite a dynamic process for enlightenment on a global basis—I call this spiritual sustainability. I've been privy to the viewpoint of the Beings of Light regarding the righteous behavior expected of us as human reflections of spiritual magnificence. In a nutshell, humankind in general is expected to do everything that I personally found so hard to do (like being consistently thoughtful, kind, and loving). For a large part of my life, all these things seemed like a huge waste of my precious time. To my old way of thinking, it was a whole lot easier to knock someone out than to waste my time politely asking them to move out of my way! But sooner or later, wisdom has a way of triumphing over even the most stubborn and foolhardy jackasses among us.

After spending over thirty years as a hospice volunteer, I've personally borne witness to the enormous grief, trauma, and bereavement that remain an ever-present companion in the human death process. In this setting, I found my spiritual power and purpose. Putting the wisdom and compassion it taught me to good use, you can explore with me the ways we can prepare for the years (and life) ahead. With the proper tools of spiritual understanding firmly in place, together we'll ride out the tidal wave of tumult rapidly threatening to engulf us. The Bible states, in Hosea, "My people are destroyed for lack of knowledge." Thousands of years ago, that was entirely true. But there is no excuse for that anymore. Not in our modern, high-tech times when so many have access to endless knowledge with a few clicks of a computer mouse. The knowledge I have gleaned as the result of my time spent in Heaven is best expressed in the Fourfold Path to Power:

The Power in Love
The Power in Belief
The Power in Choice
The Power in Prayer

ELEVEN

The Power in Love

It is love, not reason that is stronger than death.

—THOMAS MANN

In an effort to define love most accurately, I turned to my trusty desktop dictionary. The first definition was: "Love is a profoundly passionate and tender affection for another person." That sounded pretty good to me, when I considered it in the proper context. But I have to tell you, since visiting the other side, I knew love to be so much more than a mere passion or affection. I beheld love as the most powerful force in the universe. In Heaven, love is not a personal sentiment; certainly, it's no romantic feeling or even a tender emotion. Throughout my journey to the Hereafter, I saw first-hand that love is a divine, living energy of unparalleled might and magnificence. As a matter of fact, for those of us in human form, love is our true state of being.

Armed with this personal experience, I scoured the dictionary to see if I could possibly find any other definition that held the slightest resemblance to the divine love I had witnessed in Heaven. But then I had to admit to myself, love is energy so expansive, so omnipotent and omnipresent, that it could only be described as a force so powerful and profound that it nearly defies definition.

The ancient Greeks had three words to convey their understanding of the different facets of love: *eros, philia,* and *agape.* Eros was the word the Greeks used to describe the sensual nature of romantic love. Friendships of a predominantly platonic nature were described as philia. And then there was agape, the exalted love of God. A display of agape was indeed rare among humans, for it denoted love given through self-sacrifice and spiritual purity.

In light of my research, I've come to the conclusion that there is no word in the English language that can appropriately express divine love's eternal perfection. Furthermore, with all the countless volumes written through the ages to delineate the meaning and purpose of love, truly never enough can possibly be written to convey the infinite reach and endless influence of this invincible power. Unconditional divine love is not something we can formulate or simulate and hoard. It is the force that moves effortlessly throughout the entire universe and through each of us when we allow ourselves to be open channels for its divine expression. I have seen that the love of the Holy Spirit is where "we live and move and have our being" (Acts 17:28).

However, this is love as it permeates Heaven. Here on Earth, love is perceived, translated and expressed in a multitude of different ways. We all have known love that was positively more than wonderful. Yet, in other instances, love had differing degrees of impact on our lives. As an example, we love the church yet priests abuse us. We dedicate our love to family, and we are used. We chose to give our heart to that special someone only to be betrayed. These painful scenarios occur at the hand of love because we, as Spirit in form, failed to maintain our state of alignment and spiritual connection to the higher spheres of perfect love as it exists in Heaven.

In viewing love from the other side of life, I know it to be a power. This must never be forgotten, for the power of love is the key by which we identify the spiritual nature of all things. Paramount to this understanding is the knowledge that as you find something you love and admire in someone else, you are actually discovering something already existing within yourself—something deeply worthy of your love and admiration. We are constantly looking to give and receive love so we might identify with, and possibly transport ourselves to, that very special place within called the sacred. Love is the divine essence within, seeking expression without. It is the sacred emanation of the Holy Spirit streaming across the universe to animate our flesh and gift our physical lives with spiritual purpose.

You know, the only thing I believe to be as important and potentially powerful as love is laughter. Laughing along with someone else—not in making fun, but as an act of bonding—is the ultimate in shared love. Laughter manifests the divine in a moment of mutual intimate connection. To laugh together is to create an energetic level of harmonious oneness, allowing us to embrace the love already existing within ourselves. When we laugh with someone, we are divine. Repeatedly in the Bible, we are reminded to be of good cheer. For when we maintain good cheer through laughter, we fortify our state of alignment with the infinite joy of the Holy Spirit. In turn, that opens wide the door of our spirituality to welcome the flow of unlimited love.

My dear friend Michael Bernard Beckwith, founder of the Agape International Spiritual Center, says, "When we are of good cheer, we are in integrity with our eternal soul." All of life is constructed with our permission and assistance to teach us lessons. When we remain receptive, loving, and of good cheer, life's lessons cease to create life's disappointments. The point is not to dwell on our disappointments and missteps in life. Instead we must be determined to become stronger and more compassionate as the result of them. Consequently, we rise above and leave the need to repeat them far behind. Thus, the most challenging life lessons are not to be considered, in any way, punishments, mistakes, or failures. In reality, with love and good cheer flooding our spirits and infusing our hearts, the likes of confusion, doubt, worry, fear, and unhappiness are no longer part of our daily experience. For when we exist in an ever-mindful state of love and good cheer, we are in truth operating from divine mind. In that holy place, confusion, doubt, worry, and fear do not and *cannot* exist.

We are counseled in the Bible, "Get wisdom, get understanding." Deeply understanding and applying the power of love, combined with the spiritual protection of good cheer, is just about the wisest thing we can do for ourselves. When we commit to cheerfulness and to life lived in the loving expectation of goodness, the world becomes a beautiful reflection of these qualities. Dedicating ourselves to being the expression of love and good cheer, in all we do, guarantees our lives will become the direct image of Heaven on Earth. Our love for one another, put into action moment by moment, and used for the highest good of all, is most assuredly our sacred path on Earth. Each of us has a unique mission to accomplish in life. That mission is predicated solely upon the gifts and talents we chose to master, for the benefit of humanity, in loyal service to Spirit. However, in all instances, no matter what the immediate or ultimate goal might be, love is the unwavering path leading to its spiritual fulfillment. Love is the power, ever present and forever willing, to perform the miraculous deeds and unexplainable happenings destined to occur along the path to our success.

Lao Tsu, the great Chinese theosophist and author of the *Tao Te Ching*, said over twenty-five hundred years ago, "The only way to do is to be." Therefore, I believe that the only way to create love is to consciously *be* love. If there is to be peace on Earth, it must begin as a seed that first takes root within each of our hearts. We must become the dutiful spiritual emissaries of celestial harmony that we volunteered to be when we came here. All we need to do is acknowledge our essence as love and diligently use it to initiate change. Together we can change this world from a paragon of fear into a paradise of peacefulness. We have the power to transform the world with our thoughts. Now we can imagine ourselves either striving to survive in an age of unpredictability or thriving in an era of triumph.

The perspectives we adopt, along with the outcomes we experience are simply a matter of free will. Victory, like poverty or boredom, is often a chosen state of mind. A man with millions of dollars in the bank can live in abject poverty, without a single friend or even a modicum of passion for his own life. Yet, we can choose to change our minds, and thereby alter our reality, anytime we desire. Therefore, choosing our thoughts with great care is imperative. Now, more than ever before, it behooves us to remain conscious of the thoughts we allow ourselves to contemplate. Love creates more love. Loving thoughts create a more loving world. We are all part of a team of volunteer warriors embroiled in the ferocious battle for the souls of all humanity. Daily battles are being waged in the name of what is righteous, honorable, and true. Righteousness is the right use of our spiritual gifts—life and love. For in life, the righteous use of love is our timeless destiny.

It's funny now, but I can remember years ago when my mother reminded me repeatedly of how she believed true love was long suffering. I never quite understood why she kept saying that to me, at least not until I was much older. Looking back, I can see all too clearly—everyone who had ever loved me had been put through a lot of suffering for long periods of time. Of course, I regret it today. But I had to travel my path, learning my unique lessons, just like every one of my companions. I do believe those long suffering days are finally over. Due to the presence of a deeply committed love in my personal life, I've matured enormously in my ability to give and receive love on all levels. I have also learned that the divine binds and connects us in one great hope. As we give something we possess, we will in turn receive something we do not possess. It is said that it is far better to give than to receive. And so it is with love.

I give myself most purely at the bedside. When I sit with people in the last days of their life, I find I invariably revive my sense of self-worth. As I see myself through the eyes of my patient, eyes filled with appreciation and trust, I know I am broadcasting to them, loud and clear, that they are loved. They are loved for no other reason than the fact they exist. Yes! They are, and therefore, God loves them! On this day, God's love sits at their bedside. What an exquisite example of agape love as it lives, breathes, and moves effortlessly through me.

Maybe hospice is not for you as it is for me, but what is important is that you know there is a calling just for you, and that you find it. When you decide you are ready to realize exactly who you are as well as the power your love possesses, I strongly recommend you give freely of your time, love, and tenderness to someone who has absolutely nothing to give you in return. I am also a staunch advocate of mentorship. If hospice isn't your thing, I urge you to help mentor a child. Share your talents and the wisdom of your heart with a child who will become a more confident, optimistic person because of it. Just as I experience divine love in helping people feel safe as they leave this world, through mentorship, you too will come to know the quiet domain inhabited by your spiritual sovereignty. This is a regal place where you can observe the presence of pure divinity within.

Although the possibility of being emotionally bruised through loving in interpersonal relationships exists, the secret to love's power still dwells in the great strength it commands. Love is not for wimps. It takes tremendous courage to embrace this divine energy. Each of us is acting as a vessel for unlimited divinity within the confines of our physical existence. And it is time we all made the conscious decision to remember this. It is also time we started to live our lives as though life on Earth depended on us. Because it does! And it depends on our understanding of the all-encompassing nature and dynamic function of love.

Please, believe me, love molds and holds the universe together. It keeps the planets suspended in place, and it causes our world to spin on its axis. In the Crystal Cities, I witnessed love as a thriving, pulsating, radiant essence. Science calls this dark matter. Consider the atom. It is the basic building block of the universe, yet it is

94.6 percent empty space. The empty space is dark matter, and quantum physicists claim the universe is suspended in it. So, for one short moment, do me a favor and imagine that the universe, complete with its billions of stars and planets, is painted on a canvas. Imagine now that this canvas is the living power we call love. Well, guess what? We are composed of the exact same molecular constitution as that love. In order to stay in a state of perfect alignment with this living power, all we have to do is give it away. The more we love, the more we are capable of loving.

Deep, conscious breathing is the best way to center ourselves in the heart of love. By taking time each day to practice deep breathing exercises, we increase our conscious attunement to our spiritual core. Moreover, meditation is mandatory in order to discover the divine peace that resides between our thoughts. Our connection to divine peace is crucial to becoming consciously united with our higher self. From this unified state of being, our life force is magnified to the degree that it becomes capable of positively charging the subtle energy grids surrounding the planet. As the grids charge, they increase their ability to receive higher celestial energies to expedite enlightenment on a global basis. Truth is always simple; all we have to do to start a chain reaction of love is institute a daily regime that includes conscious breathing and meditation. That's so easy even I can't forget to do it!

Gratitude is another vital component in this mix. We must make a point every day of counting our blessings.

Living in a state of constant gratitude manifests more to be grateful for. Love and appreciation blended together compose harmony in the key of life, which can syncopate the collective heart and soul of all humankind into a mesmerizing rhythm of spiritual euphoria. Prayers of appreciation also need to be a part of our daily spiritual practices. These prayers are our first line of defense against negative change, not simply something we surrender to as a last desperate resort. The universe is constantly in a state of self-correction in order to maintain balance. If we offer prayers of gratitude and appreciation each day, we assist the universe immeasurably in maintaining balance. Affirmations are highly effective in maintaining balance, as they attract the positive energy needed to improve our lives. Mentally affirming the good in any situation is a potent way to charge the perfect outcome. An affirmation of love or appreciation repeated throughout the day conditions the body and soul to effortlessly receive gifts from the Divine Spirit. It is an essential part of our spiritual work.

Through prayer, we talk to Spirit; through meditation, we listen; and through affirmation, we create a means by which we can more easily receive higher energies of inspiration, abundance, and beauty.

Think about it, don't the majority of us long to live in a world where splendor, compassion, and balance reign? Don't we want to reestablish a universal value system that honors the preciousness of all life? Don't we all dream of recapturing a magic in living that embraces the spiritual essence of our physical selves? I know such dreams are possible. Love is alive, everybody. And love is not only our hope for the future; it is our destiny. It is the Holy Spirit in action. *A Course in Miracles* tells us, "There is only one of us here." In other words, we are all one. If we believe this to be true, then this power and presence brings with it the way, the truth, and the light. For the sake of our entire global family, let us dedicate ourselves with spiritual integrity to the divine responsibility of living with, and in the consciousness of, divine love from this day forward.

TWELVE

The Power in Choice

It's not a matter of can or cannot; it's a matter of will or will not. Everything is a choice; choose wisely.

—Ed Hillenbrand

Life is a matter of choice. Everything we manifest in our day-to-day lives is the direct result of our choices along the way, from one experience to the next. Each choice automatically creates a consequence; therefore, our choices need to be made from the perspective of an in-depth, conscious look at all the options at hand. Yet, let us not forget that even our perspectives originate from our conscious choices. Great consideration is necessary, for tremendous responsibility is inherent in the major life choices we make. From our choices our quality of life manifests. From our choices, other people's lives are influenced, for better or worse, as well. In the end, we are the choices we make in life.

I learned all this from being the master of bad choices. For me, good choices were not always easy to make. Prior to my first near-death experience, I rarely, if ever, considered the consequences of my behavior before I acted. My philosophy was: I'd rather seek forgiveness, than ask permission. But my two panoramic life reviews were responsible for giving me a whole new slant on this. I learned in Heaven that we have a human responsibility to be spiritual in nature, and a spiritual responsibility to be human. It is at the precise juncture of these two realities that the component of free will rises to the surface of our lives to grant us the privilege of choice.

Within one dimension spiritual beings have a human experience and have been given the freedom to construct, modify, and revamp reality, based solely on their preferences. However, this choice making is a serious challenge, one that requires great patience and wisdom from us. It also requires a clear vision of our intent. Sparkling clarity is vital if we are to use the power of choice in a manner that guarantees outcomes for the greatest good of all concerned. Whether our intent is the salvation of the world like the valiant efforts of a Mother Teresa or just personal happiness, during the panoramic life review, we will see clearly the real intentions behind our choices. We will recognize ourselves as spiritual beings attempting to create opportunities for optimum soul growth. Hopefully, a by-product of that growth is a better, safer, and more loving world.

Choosing to remain cheerful and optimistic in the new millennium is a tad tricky. Yet, it is in our best interest to strive fearlessly in that direction despite the bombardment of negativity we are forced to muddle through each day. Television, radio, and the Internet all broadcast the same bad news of our disparaging world conditions around the clock. Unless we choose to find a buffer to it all, we share the potential for falling prey to the media's oppressive parade of one human atrocity following another. Wars and rumors of wars abound. Entrenched in a culture of violence and mayhem, we must make the conscious choice not to get caught up in the hypnotic hype of hopelessness. We must stand strong in our conviction that life is a joyous opportunity to experience its awesome wonderment and adventure. Choosing to be happy in a world filled with sadness is no short order. Yet, it is our surest route to victory. We must keep our eye on the way we want things to be, the changes that need to be made, and not on the misery and pitiful conditions we hope to eradicate.

Life was never meant to be a struggle. No matter how much global struggling we witness on the nightly news, this is not the destiny of humanity. We cannot find solutions to the ills of our world by choosing to become part of the problem. When we indulge in acts of indifference or retaliation, when we choose to indulge in thoughts of violence and judgment, we become part of the problem. A component of everyone's divine mission on Earth is to choose to be pure of heart and to remain true to the cause of righteousness. Holding steadfast to the promise of goodness and the principles of truth in the face of adversity is the sacred path of the spiritual hero. In being given the power of choice, we are also given the gift of imagination. Conscious and constant application of this potent combination guarantees that we become the master of our fate. As we exercise our power of choice through the magic of imagination, we perfect our skill as sole architect of the personal world we wish to design. Our lives will be as beautiful and abundant as we choose to create them.

This principle operates on many dimensions simultaneously. When looking at something as simple as what we chose to wear on a particular day and why we chose to wear it, the greater of the two is *why.* And we must keep it that way, for the reason we do something is far more important than the act itself. The intention motivating us to take action determines the spiritual effectiveness of that action. I am not stating here that the end justifies the means. I am approaching this subject from the highest moral and spiritual ground.

When the intention is pure, not meant to be harmful in any way, the *result* of an action will always be measured before the action is. It has been said that a soul of divine character considers well the results of all actions long before they are taken. As spiritual beings in physical form, we all possess a divine character, and the choice is up to us whether or not we implement it in our daily lives. The ability to reside in consciousness of divinity, where such a character is openly displayed, is the true mark of an evolved soul. Every one of us in the world today has the ability to reside in divine consciousness. Once again, it is simply a matter of choice. Life is a creative activity. In other words, your life is in a perpetual state of self-creation.

I've been shown that everything will continue to move at a much faster rate over the next four to six years. All of us are aware of the Internet and how rapidly things are transmitted through cyberspace. From money transfers to the transmittal of technological knowledge, everything is moving at an accelerated speed. More than ever the expression *Knowledge is power* rings true. Due to our unprecedented velocity in the transference of knowledge, each choice we make now stands ready to be immediately applicable. Our imagination and perception manifest our reality. What we choose to believe becomes the mirror of our existence. Harnessing the power of our imagination then becomes a driving force, giving us the ability to rise above the everyday events that appear to control our lives.

Therefore, our experience of life is limited only by the cap we place upon our imagination. This brings the point full circle—to being responsible. As the saying goes, be careful what you wish for; you might just get it. Or better yet, be careful what your imagination chooses to believe in, for you will create it. Richard Bach said it eloquently in his inspirational masterpiece, *Illusions: The Adventures of a Reluctant Messiah:* "Imagine the universe as beautiful and just and perfect.

Then be sure of one thing: The Is has imagined it quite a bit better than you have."

As we go forth in our lives, let us not forget this. Life is the sum of our choices, predicated upon the responsibilities we willingly take on. This allows our imagination to seek the greatest possible variety of options available in the fulfillment of our spiritual destiny. I believe we cannot fail to achieve our spiritual missions on Earth. However, the means of this achievement are strictly commanded by the power of our choices. We all can share the destiny of a life well lived, a life with no regret. To live a life of excellence is to personify virtue in action. Therefore, it is our divine responsibility to choose wisely based on the counsel of a virtuous heart.

THIRTEEN

The Power in Belief

Faith is the substance of things hoped for,
the evidence of things not seen.

—HEBREWS 11:1

As children, most of us are taught to have faith in myriad things we can not see. From God to Santa Claus, and from the Tooth Fairy to life after death, we are trained to place great belief in the unseen. Unfortunately, many of the things we hold sacred in childhood do not endure with time. And so, as the years pass, we often go from true faith to a belief system based on science and mechanical thinking. Inevitably, the time arrives when as adults we begin to examine our spiritual or religious perspectives. At this juncture, we reassess what we really believe in. In turn, those beliefs become the new foundation of a more practical faith.

In our quest for an enduring spiritual truth, we may be tempted to discard our devoted allegiance to unquestioning faith. I suppose this is only natural as part of either the human maturation process or the soul's yearning for completion. Yet, the complete loss of faith can create a devastating void in our lives. Staunch belief in something greater than ourselves is an essential building block in the construction of a personal reality. If you believe in miracles, you will see them manifest in your life. If you believe in true love, eventually it will come along. Much the same, a religious faith in an unseen power, in many instances, is only a starting point. Prior to my near-death experiences, I had a lot of problems when it came down to the matter of faith—my personal faith . . .

Our Judeo-Christian heritage has taught us that we can never get into Heaven without faith in an omnipotent spirit. But, guess what? I did! After dying, as the fatal aftermath of being struck in the head by lightning, I went straight through the tunnel, into the light, and on to the Crystal Cities. I did all this without any spiritual or religious belief system in place. In fact, the only holy trinity I dared to believe in, at the time, was me, myself, and I. Yet, right after the first near-death experience, I concluded that a belief based on experience offered a much broader perspective than an adherence to any faith. For years, I thought that for those of us who have actually lifted out of our bodies, floated above them, and witnessed the separation process between the physical and spiritual self, faith was unnecessary. The vast majority of near-death survivors had been to the mountaintop. We *knew* the truth that sets us free. And I wholeheartedly contended we no longer had need of faith. We experienced life after death, a life more brilliant and sparkling than anything we have ever known on this side.

Through the years, however, my frequent reveries of Heaven's glory have persuaded me to expand my somewhat arrogant outlook. As a matter of fact, what I believe as a result of my celestial experiences has given way to a grand system of faith. It is a faith founded upon the endless possibilities the universe holds in store for each one of us. In addition, I have acquired a deeper faith in humankind's unlimited potential for spiritual evolution. You know, looking back, I am surprised that it took me so long to catch on. It was obvious to me that the journey of transition from here to the Hereafter had to have been meticulously designed by someone or something that loves us very much, because the moment we arrive on the other side, we are infused with a spiritual understanding of our eternal oneness with the infinite nature of divine love.

Through the filter of my newfound perspective on faith, I look at life today from a completely different point of view. More often than not, I now tend to view life as though I were still in Heaven, looking back at myself here. I keep that perspective always in the forefront of my mind to help me maintain a strong cognitive link between my higher self in spirit and my earthly form in the physical realm. I don't worry about having faith in Spirit. From this perspective, I worry about Spirit having faith in me. Each day, and in every circumstance that crosses my path, I am aware of my need to create trust in the way Spirit sees me handle myself. If Spirit knows I can be counted on to be honorable, faithful to my word, and compassionate in my behavior, then Spirit has a reason to trust me. That trust then inspires faith—Spirit's faith in me. The more faith Spirit has in me, the more spiritual keys I am given to unlock the secrets of the light.

The very same is true for you and everyone you know. As we instill in Spirit the faith that we are spiritually responsible, standing ever ready to be accountable in our daily lives, Spirit reveals the divine plan to us. Our capacity for spiritual responsibility, however, is founded on our ability to believe in our inherent goodness as well as our intrinsic greatness. True faith in Spirit is best expressed through a genuine belief in your fellow man. When we make the choice to develop a belief in the inherent goodness of all mankind, we are given ample opportunities to strengthen that faith. Personally, I have unshakable faith in humanity's desire—and Heaven's promise to assist us—to attain our spiritual perfection. My faith in this increases celestial faith because when the Holy Spirit is impressed with our knowledge and the practical application of our inner qualities of compassion and love, we are blessed beyond measure.

A belief in kindness that creates an ongoing stream of good deeds on behalf of others is a energy capable of reverberating far into Heaven. What you do in this life, based on a belief in the love and perfection of eternal life, decides the quality of life awaiting you in the Hereafter. Belief in compassion, love, and thoughtfulness perpetuates more of the same. As each of us becomes saturated in this realization and puts into daily practice the art of love and compassion, the world is a brighter, better place for all of us. By having true faith in one another, soul to soul and heart to heart, our mutual energies create an immutable frequency of magnified hope, strong enough to shield and safeguard our future. By believing in one another, each and every one of us literally becomes *the difference that God makes.*

FOURTEEN

The Power in Prayer

Prayer is back. After sitting on the sidelines for most of this century, prayer is moving towards center stage in modern medicine.

—LARRY DOSSEY, M.D.

As a child growing up in South Carolina, prayer played a major part in my family's daily life. I vividly remember holding hands with my siblings while my father said the blessing before each meal. Sitting beside our beds, Mother or Daddy led our bedtime prayers before tucking us in each night. In church every Sunday, I recited the Lord's Prayer by heart. My heritage is graced with true grit and the strength of genuine faith. Above all else, prayer is an integral function of the holy Southern tradition. In my early years, the family endured times of hardship resulting in loss. Through heartfelt prayer, we not only survived it all, but we did it with our sense of joy and celebration still intact.

Yet, my early religious training left me feeling somewhat empty. Unable to share in my parents' theistic exuberance, I often questioned my worth. As the years passed, I grew into manhood and slowly but surely parted company with the practice of prayer learned in childhood. Today, I clearly recognize how the seeds of my cynicism were deeply rooted in my preadolescent church programming.

One particular memory refuses to fade with the ensuing years. Because I was a little hard to handle in my youth, on the weekends Mother made me attend disciplinary Bible study classes. For wayward boys like me, too much free time spelled disaster, so my parents made sure I was occupied 24-7. During one of the Saturday afternoon discussion sessions, the instructor gave a lecture on the age of the known world. When he told us the world was roughly six thousand years old, I questioned him in earnest about how dinosaurs lived from 230 million years ago to sixty-five million years ago. With undeniable wrath welling up in him, he responded that dinosaurs were a despicable lie invented by pagans for Satan's purposes!

Little did he know that I had recently visited the Natural History exhibit in Charleston, where I spent the entire afternoon marveling at the life-size reconstructed skeleton of a Tyrannosaurus Rex. It was far too late for him to try to run such religious rhetoric off on me. With archaeological artifacts to support me, from that point on organized religion and I were embattled. Moreover, I repeatedly witnessed from a young age that God-fearing people basically prayed to: (a) keep something from happening, (b) get out of something, or (c) get through something we couldn't keep from happening.

In my adult quest for what I believed to be a more accessible spirituality, I traded in ritualistic prayer for the more popular practice of positive thinking. But one thing I never renounced is the blessing before my every meal. You see, for me not to express gratitude for something placed right in front of me constitutes outright stupidity.

After my near-death experiences, my zeal to develop a direct connection to my inner divinity swelled as the distance between organized religion and me broadened. Despite my deeply inspired pursuit of sacred knowledge, I continued to overlook one solid cornerstone of spiritual manifestation—prayer. I had given lip service to it, but I didn't truly appreciate the grounding effect it had on my life. By 1997, that was all set to change, due to what I now know to be divine intervention.

While I lay in the hospital with my life in the balance, Art Bell went on the air one night to ask his listening audience to pray for me. In my state of semiconsciousness, I had no idea this was taking place. Yet, within hours of Art's worldwide broadcast prayer request, I was startled into a state of complete awareness by the overpowering presence of a shimmering silver-blue energy encircling my hospital bed. I found out what happened many days later. Piecing things together in a timeline, I realized it only took a few hours of ongoing prayer from legions of people around the world to create a healing force of tremendous magnitude. Later I was informed that in response to Art's plea, over five million people (individuals as well as spiritual organizations and churches), in more countries than you can count, had logged onto his and other Web sites. All of these concerned, loving souls were praying for my recovery. This staggering thought humbles me to this very day.

During the entire first week of my hospital confinement, I suffered such relentless, excruciating pain as a result of the hematomas that I plunged into the abyss of despair. I contemplated death over the life of agony I was being forced to endure. Having acquired a new appreciation for it, my mind would often wander through assorted images of assisted suicide. Miraculously though, once the prayers of the caring millions started going out into the ethers on my behalf, everything shifted. Initially, my deepest concern was that, due to the endless pain, I would be incapable of mentally putting my life into manageable order.

However, in the aftermath of the mass prayer vigil, whenever I surfaced consciously and attempted to compartmentalize my thoughts, the silver-blue energy appeared in the room and the pain eased. It was an astounding realization, almost too incredible to believe. I was not privy to the details. All I knew for sure was that in the midst of the silver-blue energy, I experienced significant relief from unbearable suffering. And at those times, that was all I needed to know. In the years since those wondrous events, I have studied scientific research on the power of prayer. More than anyone I know, Dr. Larry Dossey has shined the brilliant light of timeless truth on the subject.

In 2000 I was invited to attend the Science and Spirituality Conference at Duke University. This symposium was cosponsored by the Rockefeller Foundation and geared toward exploring the effectiveness of alternative and complementary modalities (including prayer) in the healing process. This mind-expanding seminar presented truly the most compelling evidence of prayer as a medical tool. In the company of such renowned scientists as Dr. Wayne Jonas, Fred Thaheld, and Dr. Marilyn Schlitz, a multitude of amazing research data was revealed. For me, the most astounding conclusion of this elite conclave was a new definition of prayer as it would be documented in research paradigm protocol models. The term *willful, conscious intent* was coined to encompass prayer's many far-reaching medical and spiritual applications. The sheer sound of this new term resonated powerfully in my spirit, opening my mind and heart to a whole new understanding of the successful use of prayer both personally and collectively. I immediately looked up these three simple words in the dictionary to discern their spiritual connotations:

- *willful:* the ability to influence through mental power
- *conscious:* a knowledgeable stream of awareness
- *intent:* concentrated attention to a specific purpose

How marvelous! Finally, I had come to the realization of the fundamental workings of this ancient secret for bridging the gap between this world and the next. In contemplating willful, conscious intent, I was struck with the knowledge that we could assume mental and moral responsibility for our lives through the creation of a higher purpose. This revelatory event further enlightened me regarding an unforgettable aspect of the panoramic life review. This is one of the most important things I gleaned on the other side. *What* we do in our lives is not as important as *why* we do it. At long last, I grasped what really matters—our *intent*, the motivating factors supporting any action taken. This is the true measure of a soul.

Armed with this knowledge, we can take control of our lives. Adding to it the implementation of prayer as willful, conscious intent, we seize the day! In so doing we purposefully turn to the universe for use of its magnificent power. Also, let's note that when positively directed, our prayers unite with the collective consciousness in order to grant the energetic support needed. As we move ever forward in our lives, it is imperative that we utilize our willful, conscious intent to influence our experiences, along with all humanity, for the better. As we assume responsibility in our human endeavor as mighty spiritual beings, may we also intend, with every breath we breathe, to improve the global human condition. Prayer gives us the ability to draw from the spiritual realms the strength and inspiration to open our minds and hearts to behold the powers of the mystical dimensions in the physical realm. Prayer, whether used in the religious context or as willful, conscious intent used scientifically, steadfastly remains the true model of the harnessed spiritual might of directed thought. But no matter what you choose to call it, it works—just put it to the test! I believe you will see, as I have, there is indeed great power in prayer.

PART 3

The Seven Lessons from Heaven

FIFTEEN

Lesson One

The true perfection of man lies not in
what he has,
but in what man is.

—Oscar Wilde

We are great, powerful, and mighty spiritual beings of light, living in a physical world with dignity, direction, and purpose.

I vividly recall how, during my first near-death experience, the Thirteen Beings of Light impressed upon me the fact we are great, powerful, and mighty spiritual beings. I want you to realize I had absolutely no problem believing this. Especially since, at the precise moment they were telling me this, my charred body was in the hospital, covered head to toe with a crisp white sheet, awaiting a trip to cold storage. However, upon my return from Heaven, I felt deeply compelled to sort out the meaning of those words for myself. What did it truly mean to be great, powerful, and mighty? I found it difficult to fathom how we poor, pitiful humans, possessing all of our obvious shortcomings, could possibly be considered anything wondrous and mighty.

Therefore, to the best of my ability, the following is my philosophical interpretation of what I have determined to be true. It is composed of fifty-seven years of my life experience as well as my transdimensional encounters. Some of this information I was aware of prior to my near-death experience . . . and some I was not. The indwelling greatness, present at the very core of each of us, is predominately measured by the amount of courage we possess. This spiritual courage expresses itself most apparently through our determination to move into a physical body. After that, it is expressed by our dedication to manifest and sustain a physical reality. It is my undaunted belief that only the greatest heroines and heroes have taken on the task of human life during these tumultuous times. The next decade on our planet will bring the most unprecedented upheavals and unpredictable changes ever known to man.

On each of my visits to the other side, I was shown that every one of us is here for the specific mission of assisting in unfolding perfection on Earth. And this will be done in accordance with universal divine will. At the same time, we have voluntarily taken it upon ourselves to complete specific life tasks devised to facilitate the most tremendous personal growth possible in the shortest amount of time. Seventy or eighty years of life on Earth may seem like forever to us, but from the perspective of an eternal being, it is not a lot of time to accomplish such a monumental feat. Thus, we have the definition of genuine greatness: attempting to achieve a seemingly impossible task in little to no time. Doesn't that sound like courage? You bet it does!

On the other hand, our true might lies in the fact that we are unique sparks of divinity. Through us, Spirit conveys its ever-expanding power and magnificence as the sum of our manifested thoughts, words, and deeds. In fact, I believe our mortal might is expressed most profoundly through our thoughts. Our spiritual potency grows in direct proportion to the way we direct our willful, conscious intent toward effecting change in our lives, the lives of others, as well as the world at large. The changes we willfully influence, whether positive or adverse, will in turn create a ripple effect throughout the entire universe. This occurs invariably because we are One. As cliché a saying as you may feel that has become, it still remains gospel.

From the smallest pebble on the beach to the infinite vastness of the cosmos, everything is eternally interconnected in the oneness of Spirit. Let this paragraph be the primer that guides you through the rest of this book (and hopefully through the rest of your life). In this fashion the spiritual system has been established, whereby we all are acting as guardians of one another. Did you ever stop to think that the divine force could be so calculating? Well, considering the fact that we cannot be anything that divinity is not, the divine force has the capacity to be clever too, right?

Believe me, what you're reading is just one of the ways it is. I assure you, my words are far from being the kind of psychobabble that has been thrown at you along the way. These are immutable truths set into motion by the All That Is. I have my own interpretation of these truths, as do all of us. This is so because our consciousness shapes the world we live in—our experience of life, death, and the world beyond. We really need to understand this: *every single thing that one of us thinks, says, or does impacts all the rest of humanity on one level or another.* Please stop for a moment to deeply absorb what you just read.

As I evolved spiritually and learned more about this, I still found it hard at times not to be selfish and self-centered. In order to use the fourfold path to power, all selfishness and self-centeredness must be relinquished. I realize the difficulty in what I am asking of you. You have no idea how arrogant I can be. At one time, I truly believed that the only opinion that mattered was mine. No kidding! For many years, I honestly thought I was the only one who mattered in any situation and regarding all issues. But I was destined to discover the truth, and so are you, for we are here to act and live in loving Oneness. The time is now to release all arrogance. Please let it go. Begin today! In metaphysical circles it is understood that when the student is ready, the teacher appears. The fact that you are reading this page indicates that you are ready to actively respond to these truths.

By diligently and conscientiously residing in the awareness of this perspective, we automatically demonstrate a spiritual insight of the Master's edict: Love thy neighbor as thyself. For what we do unto others, we most assuredly do unto ourselves. Trust me when I say that the Golden Rule is alive and well on the other side. This is just part of the basics. Our true might is a tangible aspect of our spiritual reality as created by our thoughts, words and, deeds, for they are destined to become energy solidified and manifested in the physical realm, ready to reveal the invincibility of our divine identities.

To live in dignity, as the Thirteen Beings of Light conveyed we do, means that we possess an elevated station in the spiritual scheme of things. We were created in the image of divinity, which implies our nobility and therein the gracious conduct we were born to display. However, many members of humanity seem to have lost their dignity, which is, quite literally, the loss of their self-respect and self-appreciation. And let me tell you, this is something on which I am an expert. When that loss of dignity occurs, we subsequently lose respect and appreciation for everything around us. Then we become spiritually insulated and emotionally isolated from one another. Here, the cycle begins.

Subsequently, we find ourselves living in a society where little acts of loving kindness are unexpected surprises. In fact, kindness from a stranger can make him criminally suspect in our minds. The media has convinced us that *out there* is not a safe place. We respond to the threat of everything, from delinquency and felony to corruption and terrorism, by cocooning. We choose to engage with the world around us, but only if we can be one step removed. Through the fiber-optic barrier of Internet chat rooms and telephone conference calls, we prefer to be technologically savvy instead of interpersonally connected. Yet, a debilitating danger lies in losing our sense of interconnection with family, friends, coworkers, and community. We are being divided in so many ways that we can no longer count them. I firmly believe we were created, in Spirit, to be interdependent, to assist, support, and encourage one another to find our purpose, seek our service, and deal with our challenges in order to make a difference in the world.

That is why I do hospice work. This type of conscious caregiving makes it impossible for me to escape the one-on-one, heart-to-heart contact we deeply deserve and desire in our lives. In my personal life, volunteering time to help others is where interdependence becomes a spiritual art form. Through volunteerism, we can break down the barriers that separate us from meaningful communion with others. Volunteerism is spiritual activism in its highest form. Whenever we give with an open-heart, without any expectation of receiving something in return, we become love and compassion in action.

In other words, we are expressing as the dignity of the divine. We are creating grace, for both ourselves and the people we are helping. More precisely, *we be-*

come grace. Webster's dictionary defines grace as: (a) the freely given, unmerited favor and love of spirit, (b) the influence or spirit of God operating in humans to regenerate or strengthen them, (c) a virtue or excellence of divine origin.

So we can clearly see that the gift we call *life* is meant to be a divine state of grace. In this state, we are expressing Spirit in order to make a difference in the lives of others. This is not wishful thinking or rhetorical lip service. Our capacity to become grace is a novel idea for us to embrace, yet it is one of our cocreative rights. Most people think only God can dispense grace, but I am here to tell you that I know we have been given divine access to it. Acts of selfless love and kindness reconnect us with grace because we are spiritual beings capturing the uniqueness of a human experience.

When our human experiences are rooted in love and compassion, our outcomes are destined to be filled with spiritual grace. From this perspective, it is easy to find our true purpose in life. As I stated earlier, during my first near-death experience, I was shown that one of the fundamental lessons every soul in human form must learn is how to use our uniqueness to assist in unfolding perfection on the earthly planes. However, some of us resist, and I have to admit, I was guilty of that. It never ceases to amaze me how many times the Beings had to kill me before I finally decided to willingly do my part. I hope your path to enlightenment is less painful.

However, if that's your choice, party on! I simply can no longer personally recommend it as a viable lifestyle or gratifying path to self-realization. In the past twenty-five years, I have traveled extensively, lecturing all over the world. Though my audiences may have spoken languages foreign to me and practiced a myriad of religious traditions, one of the most often asked questions in any group remains the same: how do I find my true purpose in life? I have concluded that, in addition to assisting in the spiritual unfolding of human perfection, we all possess two essential core purposes in life. These purposes are the keystones of the next two secrets: You chose to come here. You were chosen to come here.

Suggested Focus

To apply the wisdom of The Seven Lessons as a mighty, powerful spiritual being, you must begin to hold sacred the value of your breath and your heartbeat. This is accomplished by taking meticulous care of your earthly form. For the better you care for yourself, the easier it will be for you to achieve your earthly mission, your heroic spiritual purpose. Being health conscious includes, but is not limited to, the following:

- 8–10 glasses of water every day
- 7–8 hours of sleep per night
- Regular exercise (at least 5 days per week)
- A nutritious diet consisting of the appropriate food for your blood type

- Time set aside each day for reflection and meditation (15 minutes, twice a day)
- Daily journaling of all important events and blessings
- Reading and rereading the selected Spiritual Reflection and Affirmation for each lesson

Reflection and Affirmation for Lesson One

MY MAGNIFICENCE

I am flawless. I am the beautiful reflection of my divine parentage. No matter what imperfection my humanness portrays, my identity as spiritual perfection is intact. I am the essence of the Great I Am.

AFFIRMATION

I am the magnificent expression of universal perfection.

SIXTEEN

Lesson Two

Have patience with everything that remains unsolved in your heart. Try to love the questions themselves, like locked rooms and like books written in a foreign language. Do not now look for the answers. They cannot now be given to you because you could not live them. It is a question of experiencing everything. At present you need to live the question. Perhaps you will gradually, without even noticing it, find yourself experiencing the answer, some distant day.

—Rainer Maria Rilke

We choose to come here as a force of one in order to make changes for the betterment of humanity, knowing we are capable of making a difference. Each of us has designed our life with as many obstacles and challenges as we could create, along with a variety of options and possibilities to overcome those same challenges.

So often I've heard the statements, "I wish I knew what my purpose in life is," or, "I would feel complete if only I knew what God wanted me to do." While it may take us decades to define our spiritual purpose, it seems to me that our personal quest for truth and meaning is synchronized to commence with puberty. At the same time our bodies begin the hormonal change of adolescence, our minds embark on spiritual expansion in accordance with this onset of maturity. The scope of our thinking process widens to include a personal search for meaning. In my opinion, one of the secrets to true happiness in life lies in our conscious quest for, and eventual attainment of, true meaning in life. This is the path to discovery, which leads to the essence of genuine bliss.

When we become aware and disciplined enough in our determination to look beyond the overt circumstances of the present, we are invariably rewarded with a deeper understanding of life's meaning. The reward is both coveted and sublime, for at the end of our conscious search for truth is soul wisdom. Our quest in life is not about whether we have a good hair day or if we land the big account. It is about the essential meaning at the core of these events that counts in the larger scheme of eternal life. Every occurrence, however agonizing or triumphant, fades to black as the significance behind it comes shines in the forefront of our spiritual viewing screen. The intrinsic value of any happening in life is the real power.

In our conscious search for meaning, we find true purpose in our everyday life, sometimes for the very first time. As we remain faithful to this habit of searching out meaning, we increase our internal integrity. Soon a certain pattern begins to make itself obvious. This pattern then aligns with our life purpose to reveal our divine destiny and sacred mission. Of course, each individual destiny is different, as there is no pat answer to this awesome riddle. Fortunately, though, as we consistently nurture ourselves and do the things we love to do, we come to more fully recognize why we chose to come here. So when we are in a quandary, questioning our divine direction, we need to stop, take a deep breath, and go back to what comes most naturally for us. You see, a painter must paint, and a surgeon is compelled to heal with steel.

Listening to our inner wisdom is the greatest gift we could ever give ourselves. We invite the presence of the Holy Spirit into our daily lives when we take time each day to listen to the divine thoughts floating in our hearts. Meditation is an excellent tool for self-realization. When this practice is firmly rooted in our lives, we can count on higher guidance at every turn. It's like being plugged into the celestial frequencies, nonstop. At those levels of awareness, our spiritual insight is keen and always at hand.

In the search of life's meaning and purpose, turning inward is largely overlooked as a modern method of finding enlightenment. Yet, very few answers to our spiritual inquiries can be found externally. It is said that inner wisdom speaks in the language of love, and I believe this is true. Our fundamental purpose in life is to discover our uniqueness and then to manifest it for the betterment of all humanity. This holds true for singers, street sweepers, and saints alike. The Good Book tells us, "Ask, and it shall be given you; seek, and ye shall find; knock, and it shall be opened unto you." If you truly want to know why you are here and exactly what Spirit wants you to do or be, *ask!* Then sit patiently—and with good cheer—in the silence of holy reverence and wait for the answer.

Whatever you chose to accomplish in this life, it will be done, for Spirit originally placed those most treasured dreams within your heart. Therefore, when you attain the curiosity and courage to ask for higher guidance, everything still held in mystery will be revealed. Discovering your mission puts you in a position of responsibility. From this point on, you must act as a torchbearer for others who still struggle with their spiritual identity. By this means, your life stands as a testament that everyone can do exactly as you have done. I think Spirit knew many of us would not choose to turn to higher guidance for answers, so it incorporated another surefire measure into human design. The angst of hardship and heartbreak is a powerful motivator for most people. Therefore, as we were allowed the creative opportunity to devise challenges in our lives, the more likely we would be to find our divine purpose.

So, if you have ever wondered how your life got so screwed up, here's the answer. *You* did it. You are the only one who created all the trouble you found yourself in year after year. There is positively no one else to blame. And here's the kicker: do you know why you stirred up so much trouble for yourself? You did it because you honestly believed there was nothing you could get yourself into that you also couldn't get yourself out of.

Your eternal nature is that of a spiritual being, while your humanness is a fleeting experience in your big picture. But as a spiritual being, you are naturally arrogant and feel the need to prove you can outmaneuver any obstacle set before you. Over time we become decreasingly enamored with struggle and pain, opting instead to work smarter, not harder. This kind of expanded thinking takes us full circle, back to the days when we first started questioning our purpose on Earth. Having lost our taste for failure, repeated mistakes, and dead-end relationships, we aspire to something more, something higher. And when we do, voilà! Spirit stands ready and willing to help.

Allow me to emphasize the fact that these hardships and devastating knocks along the way also play an important role in our personal evolution. They serve to make us more resilient and mature. We've all heard that what doesn't kill you only makes you stronger. So let us all pray that we know more at sunset than we knew at dawn. Remember, you can never become a great sea captain if you sail only calm seas.

Purposefully preparing yourself to receive spiritual insight guarantees it will manifest. Take time each day to ask to be shown your life's true mission. Open your heart, and keep it open, to receive. Inspiration will come and, with it, you will know a peace that surpasses all understanding. I was shown in Heaven that only the best and the brightest souls have chosen to come into the physical realm at this point in history. We are the bravest of the brave. Born of great courage and might, we have taken upon our shoulders a tremendous task of priceless consequence.

Collectively, we have come to save the world from the destructive ramifications of humanity's inferior nature of greed, violence, and the need to conquer. In this battle of mores, love must be your shield and your sword. The way you wield your love constitutes, and literally determines, the fulfillment of your personal mission. Finding and fulfilling your unique mission guarantees success. In the eyes of Spirit, you are already a glorified champion simply because you chose to be here in this extraordinary era of human evolution.

Reflection and Affirmation for Lesson Two

CHOICE

As I entered this physical realm, I brought with me two things: my personal mission and free will. I am free to choose happiness, enlightenment, and higher thought, no matter what circumstances challenge me this day.

AFFIRMATION

I protect and encourage my right to choose the best in myself, for myself, always.

SEVENTEEN

Lesson Three

Everybody can be great . . . because anybody can serve. You don't have to have a college degree to serve. You don't have to make your subject and verb agree to serve. You only need a heart full of grace. A soul generated by love.

—Martin Luther King Jr.

We were chosen to come here. This means that a great and infinitely loving force trusts and believes we will achieve the goal we were sent to accomplish. We will do this in the name of what we hold divine. I firmly believe that very few of us will ever fail this part of the life mission.

This third lesson from Heaven might upset the apple cart for some folks. With an understanding of how and why we were chosen to come here, it becomes evident that the religious concept of humanity being born in sin can simply no longer hold water. Once again, it reminds me of all that "streets of Heaven are paved in gold" business. I just have to tell you, I know it is not true. Why? I have traveled to Heaven and seen everything on the other side of life. I returned with an abiding knowledge of just how much we are loved and revered over there. Not once was I ever made to feel that I, along with humanity, was sinful or somehow inherently flawed. Never, not even for a single moment, did I sense that we were being viewed as anything less than mighty spiritual beings traveling collectively along a path of growth. To reiterate, the Thirteen Beings enlightened me to the fact that we are indeed considered to be the most courageous heroines and heroes by those who are watching us from the celestial realms.

The truth is we have actually been entrusted with the fate of the world. Please understand this: we bear the great responsibility of transforming the spiritual reality of this physical realm. Why in the world do you think they would have considered giving this enormous task to a bunch of sinners?

Unfortunately, no matter how ridiculous the concept of original sin sounds to me, the idea did catch on. And sadly this born-in-sin business has contributed to our losing sight of our identity, our dignity, and our divinity. Let me tell you how this whole religious misconception began. The notion of us being born in sin did not come into existence until the Edict of Milan, today known as the Nicene Creed. The year was 325 A.D., and the place was Constantinople. The main instigators were Emperor Constantine of the Holy Roman Empire and the reigning bishop of Rome. Together, these two men conspired to unite the Roman Empire under one official religious perspective. Using a small Hebrew cult known as Christians as their vehicle, they turned the cult's sacred spiritual beliefs into the mandated religion of the Roman Empire. With the creative assistance of the bishop, Emperor Constantine called for a council of 317 bishops to meet in Nicea, located near what we now call Istanbul.

Under the direction of Constantine, the council literally invented a version of the Word of God which more fully meshed with their personal viewpoints and political aspirations. With this revised version of Christianity, Emperor Constantine coerced the empire to embrace the concept of being born in sin, which made all adherents of the new religion subservient to his personal control. This was coupled with the indoctrination of what was about to become the Holy Roman Church. In order to keep true believers captive in a state of strict obedience, the early Church relentlessly preached the presence of Hell. For those who did not conform to the chokehold that religion placed on their lives and finances, a fiery inferno awaited. Without a doubt, existing under that kind of fear and dominance negatively affected their sanity as well. As a result, it created a perversion of goodwill toward other members of the community. The human heart cannot be cruelly coerced by fear of threat, or worse yet, loss of eternal life and remain unscathed.

The maladies that afflicted people during the times of the Roman Empire are quite similar to those affecting modern society today. Due to overbearing governmental controls, citizens hold one another suspect and tend to cocoon instead of reaching out to embrace a sense of community. Open-heartedness is gravely diminished when people are forced to live in an atmosphere of manipulation through demagoguery. Christ said he came to teach a new way and asked that we do unto others as we would have them do unto us. Above all else he asked us to love one another. These are the spiritual principles each of us has been chosen to express in the world at this time. Just like the Master, we did not come to create a new religion. We were chosen to come here to more fully reflect and express the divine principles of love, compassion, kindness, and good cheer already innate in our nature.

These qualities are paramount to the continuation of life on Earth. Now more than ever before, the world is in need of the positive influence of a humanity living and loving in an elevated consciousness. With terrorism and wars continuing to rage, the possibility of global annihilation increases in direct proportion. The purpose of all people chosen to come here at this point in history is to save the planet from an ignorance of spiritual law that could lead to our final destruction. So, in summary, we were chosen to come here to teach the lessons of a practical adherence to spiritual principles based on celestial consciousness.

The way we are living now isn't working. If we are to endure, a change in the way we think is necessary. Moreover, the fact that we were chosen to make the journey into the physical realm during these times means there was absolutely no one else who could do our specific jobs, on this level of life, better than we could. It has been scientifically proven that the chances of another person exactly like you living on Earth are approximately one in sixteen billion. Today, the Earth's population is just over six billion, which scientifically confirms the fact that no one can do what you do, exactly as you can do it! You were created as an exquisitely unique and utterly special human being. You are literally one of a kind. This remains constant whether you have realized the truth of your life's purpose or not.

Believe me, you are on a divine mission. Regardless of what your official job title might be at this moment, you have been sent here to act as a beacon of light to help others find their way. Looking from the spiritual realms back into this mental/physical world is much like gazing into a mirror, for the image our lives appear to represent is reversed. Often it is just the opposite of what we think it would be. For example, how many times have we attempted to impress the world with enormous acts of heroism, self-sacrifice, or generosity? We believe that magnanimous behavior will get Spirit's attention in a big way, winning us major brownie points in Heaven. The truth is, the system does not work that way. From Spirit's vantage point, a very different honor system takes priority. The small acts of spontaneous love and compassion are most highly treasured in the eyes of Spirit. Likewise, what you may think of as your greatest weaknesses may, in fact, be your truest strengths as seen from the other side. Judge not—not even yourself—for you cannot completely know the ultimate plan for your life. Judgment is not necessary for the achievement of that plan either. What is necessary is that you live, at all times, conscious of love. From that sacred place, you will undoubtedly be led to the perfect situations and relationships in need of your compassion and healing.

It is imperative that you do not forget this, the third lesson from Heaven. This secret of the light is tantamount to your genuine success as a light bearer in your adventure through the misty twilight of this era on Earth. To ensure that the world survives in grace, you were chosen to come here. Because there is no one exactly like you, no one can take your place or fulfill the demands of your destiny. And destiny demands your conscious commitment to a life of excellence, lived in loving service to Holy Spirit. When love alone acts as your compass, your life is always right on course.

Reflection and Affirmation for Lesson Three

PREPARING MYSELF AS A VESSEL

Throughout lifetime after lifetime, my soul urge has been toward the attainment of truth and spiritual enlightenment. Finding the answers to the mysteries of life has been my driving force. Preparing myself as a vessel of goodness has been my holy mission.

AFFIRMATION

It is my destiny to serve as a vessel for the light of love.

EIGHTEEN

Lesson Four

The person born with a talent they are meant to use will find their greatest happiness in using it.

—JOHANN WOLFGANG VON GOETHE

WE ALL POSSESS an abundance of gifts and talents. One vital aspect of living this life is discovering and developing the wisest use of these talents in order to produce the greatest potential for good.

In the traumatic aftermaths of my three near-death experiences, as you can well imagine, I spent many hours contemplating every detail of my escapades in the Hereafter. This was especially true in the two years directly following the lightning strike. Initially, I was physically unable to lift myself out of bed. And after recuperating for many months, when I finally mustered the strength to leave my bed, I usually woke to find myself face down on the floor, having passed out cold. I am genuinely without a means to convey the extent of the physical pain I suffered, along with the tremendous emotional struggle I endured in order to return to my life after the first death experience.

During those first two years, I had nothing but time on my hands to sift through my celestial experience, again and again, with a fine-tooth comb. Trust me, I searched repeatedly through every nuance of wisdom telepathically shared by the Beings of Light, and every tidbit of evidence, desperately hoping to gain an understanding of the mind-boggling events I bore witness to on the other side. In retrospect, based on what I saw and heard in Heaven, I believe every soul is born into this world with a talent and a task, a memory and a mission; everyone arrives with a gift and a goal. These talents and tasks are Spirit-given. They are our passports into the physical realm, for when properly expressed, they assure the continuation and evolution of the physical world.

Some of us are born to be scientists, while others may be inventors or even healers. Some among us are born to unveil the secrets of the universe, develop higher technology, or perhaps, engineer ways of living longer, healthier lives. Others of us may have been born with the blessed gift of music or art, writing or speaking. But one thing is for sure, through each of our efforts, as we put our talents to the wisest use, the world is made a more beautiful and harmonious place to be. You know, as I think about it, most of us are born with an abundance of gifts. One vital aspect of our physical journey is to consciously search for and fully develop these spiritually granted gifts for the purpose of manifesting the greatest potential for the good of all. Within this particular truth lies the greatest power. It is easily accessible for your spiritual evolution. Comprehension and conscientious implementation of this truth literally put you in control of your life and destiny. They will also ensure that you reclaim yourself as a mighty and powerful spiritual being.

After years of contemplation, I have figured out how this system operates. See, we are endowed with certain talents so that, in using them, not only do we find true happiness and meaning in life, but we also authentically help others. Here again we can see firsthand how everything in the universe is interrelated and interconnected, right down to the gifts we came into this world possessing. But the hitch in all of this is, how do you discover exactly what your personal gifts are? *More than anything else, this process of discovery will hinge upon what you most sincerely want to accomplish in this lifetime.* Do you want to be a healer? Do you want to teach the eternal truths? Do you want to compose beautiful music? Do you dream of finding a cure for cancer? Or maybe connecting with plants or animals is your passion.

Without being privy to your innermost desires, I can still guarantee you this: your dreams and aspirations are the real clues that will lead to your spiritual mission in this lifetime, a mission that can only be accomplished through the practical application of your inherent talents. In other words, whatever brings the greatest joy to your heart is the exact direction your life must follow. Yet, even with this in mind, we need to be flexible, for as the Beings of Light told me, nothing is cast in stone.

Let me give you an example. Let's say you have dreamed your entire life of being a healer. So you study hard to be accepted into medical school but, despite your best efforts, you don't make it. Do you then simply give up the idea and abandon the dream? Most definitely not! This is a sacred quandary. You have happened upon a moment of destiny, and you must use moments of destiny to stir the soul of your creativity. Friedrich Nietzsche, a nineteenth-century German philosopher, once said, "You need chaos in your soul to give birth to a dancing star."

Your dreams and aspirations are flaming torch lights leading the way to your dancing star. Perhaps, it is only the configuration of the route to your star that needs tweaking. Becoming a healer is not limited to being an internist or a surgeon. The art of healing takes many forms and includes myriad styles and traditions. Could it be that Spirit needs a few more hypnotherapists or Reiki masters and you are one of them? Or consider this: What if you aspire to become a singer with millions of adoring fans, yet it seems the only singing gig you can get is with the church choir during Sunday service? While you might experience disappointment, Spirit may view this as the most positive and meaningful expression of your talent *for the time being*. Who knows what the future may hold? None of us can predict with 100 percent accuracy what is to come.

True joy in life comes from living in the moment while learning to bloom where you are planted, even if you think you're not where you should be. And it is precisely at this kind of juncture that I would urge you to return to the path of power. The spiritual pillars supporting the fourfold path of power—prayer, belief, choice, and love—will assist you in cultivating patience as you wait to receive clarity regarding the precise manner in which Spirit desires your talents to be manifested.

In prayer, be specific, giving thanks for spiritual guidance to direct you to the perfect vocation in order to develop your gift of healing. In belief, know that your true joy will be found in the expression of your talent. Choose to show due diligence as you allow the process to naturally unfold in Spirit's timing. In all ways, let love be your guide. In the light of love, you will move ever closer to the realization of your sweetest dreams. Of course, it's best to consciously travel the path of power every day, in spite of whatever challenging circumstances you find yourself in, for no matter what situation might confront you, remaining allegiant to the path of power will invariably lead you to your highest good and closer to birthing your very own dancing star. That is your gift *from* Spirit. Living your sweetest dream is your gift *to* Spirit.

Reflection and Affirmation for Lesson Four

SPIRITUAL GIFTS

The Holy Spirit has blessed me with abundant gifts and talents. It is part of my Earth assignment to recognize, strengthen, and utilize these gifts to my highest potential, for the good of all.

AFFIRMATION

I devote the use of my divinely given gifts and talents to the upliftment of humankind.

NINETEEN

Lesson Five

Life lived for tomorrow will always be just a day away from being realized.

—LEO BUSCAGLIA

WE CHOSE TO BE alive at this time, at this place, and at this point in history. Never before have we had such a glorious opportunity to display our individual power and presence. Think of this: When someone gives us a gift, it is called a present. The present we have been given today is the Now. This is the exact moment to truly know that we can make a difference.

By this point, I am sure you have begun to grasp the importance of seeing yourself as a mighty, powerful spiritual being who chose, and was chosen, to come here with an abundance of talents. Undoubtedly, for the sake of flexing your spiritual muscles, you selected the perfect point in the history of our world to strut your cosmic prowess. No one else could possibly do what you have come to do in a more positive and powerful way. The acceptance of this fact is one of the greatest gifts you will ever give yourself. Rather than struggling to discover your spiritual identity and your true purpose in life, take a breath and relax. Do not strive; simply *accept* your divine destiny.

Wherever you are, right this moment, is exactly where you are meant to be. And wherever you are, whatever you are doing, you are on a sacred mission. Of this I am absolutely certain, for I was shown this truth in the Crystal Cities. The Thirteen Beings of Light instructed me about the importance of cherishing each instant in time, for every moment holds purpose and meaning. Via their divine teachings, I was deeply impressed with the significance of being present in the moment, of living in the Now. There is no greater power than the present. It is all we truly possess. There is only now. Yesterday is nothing but a memory, and tomorrow is only a dream. In this instant, we have the one great opportunity to manifest our love, power, and mentorship of the young and old. Remember, as Richard Bach wrote in *Illusions*, "You teach best what you most need to learn."

In a divinely engineered scenario, I was taught how fleeting life can be. At the age of twenty-five, it only took an instant for my life to be taken from me, in every sense. As a consequence of the lightning, I lost my ability to walk and feed myself. Hence, I was stripped of my power to earn a livelihood and ultimately, my relationship was destroyed. In reality, one and a half seconds in time cost me everything I held dear. One and a half seconds, that's how long it took for the lightning to kill me. Before that, I never gave the time of day to any thoughts on death and dying (except when I gave military targets the maximum opportunity to die for what they believed in). Staunch in my conviction that I was invincible, I figured I would live forever—just like most of us think we will. My entire existence revolved around what I wanted and the easiest way I could possibly attain it. I was all about me, me, and more me! Well, as self-absorbed as I enjoyed being, at least I was living in the Now!

Yet, on a more serious note, in the real sense of the Now, time and space are nonexistent. In Heaven, what we perceive to be the past, present, and future all exist simultaneously. The panoramic life review showed me that it is all a matter of focus. When we stay focused in the present moment, we enter what I like to call the safety of the Now. The best way I know to achieve the true safety inherent in this zone is to either teach or learn one of the four pillars of the fourfold path. For example, in learning and teaching the power in prayer, we help to augment humanity's ability to cocreate meaningful and positive outcomes in life through willful, conscious intent.

As we remain deliberate in our determination to make wise choices, we exhibit our spiritual nature of goodness and righteousness. As a result, it is easy to see how remaining in alignment with the pillars of prayer and love keeps us focused on the Now. When we pray and believe our prayer will be answered, there is no need for concern. We don't have to worry about what was or what will be. Our prayer for the perfect outcome transports us to a place of perfect peace, in the present moment. Staying focused in that peace, moment by moment, is the challenge—but it is also the reward.

The same is true of the power in love. When we live in a consciousness of love, there is no past or future. For love conquers all things. It forgives all things. Divine love is timeless; true love is enduring. In the fourfold path of power, the pillar of love stands supreme. The expression of divine love has never been needed on Earth more than it is needed now. As Lesson Five states, never before have we had such a glorious opportunity to display our individual power and presence. At this time (the dawning of the new millennium), at this place (Earth—the jewel of the galaxy), and at this point in history (the birthing of a new spiritual paradigm), we have been called into the ultimate spiritual battle. The stakes are high, and we possess the courage to take the necessary risks, for we are engaged in the mighty battle for the souls of men. Our battleground is the landscape for balancing polarity, and our victory will defeat the belief in duality.

Much of the spiritual work accomplished over the past two thousand years was carried out in the spirit of finding harmony within polarity. In order to achieve the level of spiritual transformation required to ascend into higher frequencies of light, we must balance polarity. This is yet another vital aspect necessary for divine wholeness. The inner integration of what we consider light and dark, yin and yang, love and fear, is indeed a major triumph in our quest for personal and spiritual development beyond the third or fourth dimensions. We are already in the transition phase, prepared to move between dimensions. What holds us back is our belief in duality, our reliance on a belief in opposites. We have been taught to see the world in black and white, good versus evil. As the end times bring about the demise of this antagonistic mindset, we will behold the predominance of darkness. This is a natural part of the transition into unity or balanced polarity. Contrary to what we have been told, the light does not need the dark to exist.

The forces of resistance and enmity will appear temporarily reinforced as they make their last stand of tyranny. However, as we eliminate the need for darkness within ourselves, we neutralize its power in the world. Our personal renunciation of fear, doubt, and negativity in our thoughts, words, and actions will in turn cause darkness in the world to diminish. Then we will watch as it gradually disappears from the face of the Earth. Through the faithful practice of the four pillars, we are assured the experience of unity over duality. This is a logical consequence of expanding our consciousness into the reality of spiritual purity. We shall see the promised fruits of our labor, for then we shall take our rightful place as the bringers of the glorious and long-awaited Age of Peace on Earth.

Do not be deceived; this *is* within our power. Stop at this precious moment in time and realize it will be *you* who truly makes the difference. Stay in the Now, and make miracles happen today. Seize the moment, and know that in this instant, you can make the difference. You might be wondering how to measure a moment. Well, I admit it's relative. A moment to us can be an isolated second, frozen somewhere in time, a mere twinkling of an eye. Or it may be an entire day. Yet, to an angel, a moment may last a million years. Regardless of how you choose to measure it, a moment is an opportunity to make your presence known. And through the adoption and application of the four pillars, you can use that one moment to make a positive difference.

In the final analysis of our lives, we will judge ourselves according to the difference we made in every event that transpired and for every person who crossed our path. That is the key to the successful conclusion of our common spiritual search for meaning. Every breath you breathe in and every breath you breathe out is an opportunity for emotional change, spiritual growth, and unfolding the divine. For we breathe in for ourselves and out for Spirit. The sacred space between the breathing in and the breathing out of our life force is where Spirit resides. In that holy and mystical pause between breaths, we find the home of our divinity and the birthplace of all miracles. As we determine to use each moment to help another, we seize the day by aligning with the power of the spiritual nature within.

Take the time to feel your might and express your gifts. In so doing, *you* become the present of the Now.

Reflection and Affirmation for Lesson Five

EXUBERANCE

In all the cosmos, there is positively nothing more thrilling than the presence of spiritual exuberance. While basking in the sparkle of this awesome quality, one is magically transported to the heavenly kingdom. To emanate spiritual exuberance is to emulate the passionate nature of the universal creative principle. In all ways, I will seek to bring exuberance to my life's tasks, my personal being, and my sacred space.

AFFIRMATION

The power and glory of Spirit is evidenced in my exuberance for life.

TWENTY

Lesson Six

Remember where you have been and know where you are going. Life is not a race, but a journey to be savored each step of the way.

—NIKITA KOLOFF

WE EXISTED in another world before we came into this one. It is beautiful. It is safe. It is all embracing. The foundation of that world is that we are loved, cherished, and considered precious.

I imagine you are reading this book because you are interested in the subject of life after death. In this chapter, I will attempt to share with you what I know about the quality and atmosphere of the life before this one. I do not possess the language to adequately describe this realm. (I speak Southern American, not English, you know.) However, to the best of my ability, I will try to convey what I have been able to discern from my experience on the other side. I believe that in order to fully grasp the amazing adventure we call life, it is important for us to take a deep, reflective look at the world we came from. The missions of greatness we have chosen to accomplish here and now are the direct result of the extraordinary spiritual clarity we gained in the world before. The length of time we inhabit this realm is decided on an individual basis. Yet, the reality of this spiritual field proves that the belief that we were created at birth is a myth.

Before we entered this world in physical form, we were held in the magnificence and protection of the divine womb. This all-embracing experience of safety is a testament to the fact that we are more familiar with this sacred location than any other place we have ever been. Within the spiritual sanctuary of the world before, we have always been held precious. We are loved and cherished. The universal womb resides beyond any dimensional reality we could possibly find conceivable at this point. It exists at the precipice of all beginnings, standing at the portal to infinity.

In our life before life, we are permeated with the energy of spiritual purity. Thereby, our souls are infused with the integrity of divine innocence. In this haven of holiness, we experience our primordial origins of life while we are bathed in the warmth of eternal truth. Subsequently, we are compelled to expand our consciousness within a framework specifically designed for the universal progression of love and compassion.

Our time spent in the cosmic womb is structured to be developmental. In this treasured environment of enlightenment, we do the arduous work of soul excavation. We probe the deepest strata of our being to bring to light our most authentic selves. In this way, we come into the realization of our powerfully intimate connection to everything else within the universe. The world before is where the blueprint of our spiritual evolution is lovingly fashioned. Every lesson necessary for our soul's growth is scrupulously planned in this divine realm.

In the world before this one, our spiritual progress is carefully measured and its advancement meticulously calculated. You know, in many ways, we remain forever connected to this realm, for I believe that when we elect to take on a physical body and life, an aspect of our eternal being stays behind in this realm. And this is one of the reasons why we can communicate with our departed loved ones as well as other dimensions and realities. We do not enter the Earth plane without a deliberate course of action. When we choose to leave the universal womb, we enter the physical realm to actively participate in the execution of the divine plan for spiritual evolution. However, once we cross the veil into the solidified energetics of the third dimension, we are granted free will. The choices we make as a result ultimately determine how we arrive at our destiny, a destiny that was decided upon in the world before. Every stage of our spiritual fate is lovingly prepared in order to assist us in the fulfillment of providence—the creation of Heaven on Earth.

Reflection and Affirmation for Lesson Six

KNOWINGNESS

I cannot rely on the appearance of the present as a means of predicting the future. When I know in my heart, unwaveringly, where I want to go and what I want to achieve, there are no challenges that can stop me.

AFFIRMATION

My knowingness tells me that my dreams were first dreamed for me, by Spirit.

TWENTY-ONE

Lesson Seven

Seeing death as the end of life is like
seeing the horizon as the end of the ocean.

—DAVID SEARLS

THERE IS A WORLD that exists after this one. In fact, it is the same world we left to come here. What we leave in the physical world is the uniqueness of our divine mark. And what we take with us into the next world is the understanding of how and why this divine mark was our unique destiny.

What I refer to as the world beyond is the place to which we will all return following the transition we call death. It is also the same world we left in order to allow our souls to be born into this life. So the world before and the world beyond are one and the same. However, one very important difference distinguishes one from the other. Actually, the difference lies in the delivery system—going in and coming out of Heaven is strictly monitored. The most elaborate and stringently operated way-stations are in place. These divine clearinghouses are meticulously maintained in collaboration with the spiritual hierarchy. You know, I have read that people who remember being born vehemently state that it's a whole lot harder to come into this world than it is to leave it. I can believe that, too!

But let's think about this: Over six billion people have left Heaven in our lifetime alone. Each of these souls ended up in the right body, in the right family, in the right country (although I have to admit that my family often wondered where I had come from and why on Earth they had to put up with me!). According to divine protocol, all souls entering Heaven from the physical realm, as well as those souls ready to ship out of Heaven, are classified and categorized according to their level of consciousness and final destination. These impeccable ports of exit and entry are located directly between the present physical reality and the cosmic territories of the beyond.

Our experience at the divine clearinghouse is like a gentle cosmic cleansing. Quite specifically, we are calibrated to the vibrational frequency of mental innocence that acts to clear our minds temporarily of all memories unnecessary in the upcoming life. One of the reasons for living within the physical plane is to learn, grow, and mature spiritually into a state of unconditional love. If we had full knowledge and memory of all that we are and everything we have ever done, the work we have to achieve on Earth now could be undermined. Our focus might be distracted by glimpses of the past, causing our present mission to suffer due to lack of attention. Therefore, the veil of forgetting is a divine dispensation. As the process of departure from the world before unfolds, we cross through the veil of forgetting to commence a pristine lifetime. However, we are born into the world with cellular memory encoded with specific triggers. Over the course of our lifetime, these sparks of spiritual intelligence prompt the soul to reclaim and develop its talents. The timing of the release of these cosmic memories is carefully calculated to occur with precision.

When the soul realizes a sufficient degree of accomplishment on one level, more soul memory is freed from the subconscious in order to propel soul growth to the next level. In the University of Life, we progress from one level or grade to the next, in succession. However, as it often happens, a student might have to repeat a grade—maybe even more than once—before perfecting any given lesson. In the heart of Spirit, it does not matter. All lessons will be learned in perfection, sooner or later. Spirit has only two opinions of our spiritual progress: good and very good. Even our mistakes serve to make us stronger and wiser. Thus, the wheel of life continues to turn until the soul completes its earthly assignment. This could take one day or one hundred years. Each soul and every mission is unique. Let me assure you that every one of us will wind up exactly where we are supposed to be at the end of our lives, no matter what. The divine plan has made allowances for every unforeseen contingency we could possibly throw at it.

The universe is wise enough to know that without having full knowledge of why we are here, we will probably screw up along the way. However, the good news is, unfolding universal perfection is an absolute given. Everything is going to turn out just the way Spirit has planned it. So we can act up and be resistant all we want; the will of Spirit will be done, regardless of what we decide to do.

Accordingly, when we arrive at the preordained moment to release our present life, we lift out of our body, observing ourselves from a vantage point far above ego and human frailty. From this elevated position of purely objective observation, our souls can behold an incredible panorama of spiritual awareness. We are granted a few moments to center ourselves before we are moved beyond this dimension. In that time, we become one with the perfection of the universe. Our consciousness instantly perceives the divine order in all things, and we experience no fear in our transition. All life is consciousness and, because we are expressing as consciousness, we are able to manifest at differing levels. As we die to the consciousness of the physical realm, we translate ourselves into the consciousness of the higher, more expansive realms.

Although there is a small, momentary pause to help us adjust to leaving this life, there is also a very large part of us that is more than eager to be liberated, for we know we are going home. Without any doubt, an aspect of our immortal soul remembers the way and relishes in the prospect of returning again. Death is not a finale. It is no more an end to life than a sixth grade graduation into junior high school is an end to your academic career. Both of these promotions are simply new beginnings, moving us into a more advanced and sophisticated learning system. Life energy (consciousness) is eternal. Nothing can destroy it and, by design, it is destined to expand eternally. Personally, I found the process of transition fascinating. After the lightning blazed through my body and exited with my soul, I lifted straight out of my body (through my chest) and hovered above the scene below. As friends tried frantically to revive my charred body, I watched from above. Bathed in an aura of peace, I was accompanied by a sense of utter detachment.

It was like being at the movies. I found the action interesting but in no way did I take it personally. After a few moments of this objective bird's-eye viewing, I intuited a spiral of light forming over my shoulder. At the very instant I gave my attention to it, I became a part of it. Inside this swirling energy, I perceived myself to be in a tunnel of sorts. In the distance, I could see a bright, brilliant, beautiful light at the end of the tunnel. When I reached the light, I encountered a Being who overwhelmed me with unconditional love. My only desire was to immerse myself in the luxurious bliss of this state of being. I could not remember ever knowing this kind of ecstatic, sensual euphoria. Once I felt it, though, I had no intention of ever leaving it. I guarantee it. Next came the panoramic life review, and I will devote the following chapter entirely to that subject.

As I have said, I was returned to my life against my will. I had a mission to accomplish, and I would not be allowed to stay in the light until it was achieved. In large part, my mission has been accomplished through my work with the dying. I have been a hospice volunteer for almost thirty years. I have accrued more than sixteen thousand hours at the bedside, and been at the bedsides of 348 people (including my mom and dad) as they took their last breaths. In many cases, I stood witness as they lifted out of their bodies and started down the tunnel.

Due to my volunteer work in hospice, and especially with our nation's veterans, I have acquired an in-depth education on what needs to be done in preparation for death. Whether it is our death we face or the death of someone we love, there are certain steps we will go through in the grieving process. This process was first formally delineated by the late Dr. Elisabeth Kübler-Ross. Elisabeth was a dear friend of mine, and she is deeply missed in this world. Indeed, she will forever remain a true hero of our time.

In 1969, Elisabeth wrote *On Death and Dying*. In this groundbreaking work, she was the first to delineate the five psychological stages of dying: denial, anger, bargaining, depression, and ultimate acceptance. Every one of us is going to lose someone we love. Every one of us is going to face our own death. With that in mind, we all need to acquaint ourselves with the five stages of dying. Beyond that, we need to minimize the additional sense of trauma and confusion that death can create. By this, I mean that we need to give it time and forethought so our families do not have to muddle through the horror of our lack of preparation. Due to the fact that transition from this world to the next can occur without notice, as responsible adults we need to make sure certain key elements are in place. Below is a checklist for you to use as a guide.

Make sure you have:

1. A personal will stipulating your final wishes (where you want to be buried or if you prefer cremation) and the bequeathing of your assets.
2. A living will delineating the medical treatments you do and do not want administered.
3. A healthcare proxy to make any medical decisions you may be physically or mentally unable to make.
4. Insurance policies and financial documents up-to-date and where they can be easily located.
5. A pre-purchased burial plot and burial insurance, if cremation is not your preference.
6. Last, and most important, a life of love and laughter to be your lasting legacy.

With this careful preparation for the inevitable, we free our minds to enjoy our lives more fully. When death becomes a reality through the loss of a loved one, closure is an issue of prime concern. By taking care of the items listed above ahead of time, we allow everyone touched by the loss to reach closure a lot more easily. It is said that while coming to terms with loss takes the average person two full years, healing from it can take much, much longer. There is no right or wrong way of dealing with grief, nor are there time limits on our sorrow. Tenderness and patience must be exercised with ourselves and one another as we go through the natural phenomenon of death and dying.

Once everything is in order, we must pay attention to whether or not we are practicing the four pillars of power—prayer, belief, choice, and love. These simple applications can make a world of difference in the quality of life we will experience through the coming years of change. As we concentrate on evolving spiritually and releasing any fear of the *life*—not death—to come, we can focus more attention on living lives filled with love and compassion. In the next chapter, I will share the wisdom of the panoramic life review and how we can prepare for it long before we experience it.

Reflection and Affirmation for Lesson Seven

REMEMBERING THE DEPARTED

There is a legacy of understanding and love left behind by each and every soul who ever walked the Earth. For no life was ever wasted. Every lesson ever mastered by a fellow soul shines light on my life. Today I take time out to remember those loved ones who have ventured beyond this realm yet have left behind a plethora of wit and wisdom from which I might draw a bit of strength and serenity.

AFFIRMATION

Today I will contemplate the lives of those departed, and I will be grateful for the priceless gifts that their lives have bequeathed me.

TWENTY-TWO

The Panoramic Life Review

I shall tell you a great secret, my friend.
Do not wait for the last judgment, it
takes place every day.

—ALBERT CAMUS

At the instant of transition from our physical bodies back into our light bodies, a silver cord attaching us to the third dimension is severed. With the severance of the cord, the soul is released and we are free again. Yet, a loving protocol is still in place to usher us through the remaining steps of our transition from the here to the Hereafter. Lifting out of the body and moving down the tunnel are effortless on our part. However, when we arrive at the light at the end of the tunnel, an entirely new game is about to begin.

As Lesson Seven states: What we leave in this world is the uniqueness of our divine mark. And what we take with us into the next world is the understanding of how and why this divine mark was our unique destiny. I know this is true, yet many souls departing this world for the Hereafter have no conscious understanding of the unique mark they possess. This is where the panoramic life review comes in. When you have a panoramic life review, you literally relive your life in a 360-degree panorama. In astonishing detail, you see everything that has ever happened. For example, you can count the number of hairs in the nose of the doctor who delivered you at birth. You can even see how many leaves were on the tree in the front yard when you were six years old and playing in the dirt. You literally relive it all.

Next, you watch your life from a second person's point of view. In our society, we are taught to be sympathetic toward others, but from the second person's point of view, you'll feel empathy, not sympathy. You literally become every person you have ever encountered. You will experience what it felt like to be that person, and you will feel the direct results of the interactions between you.

When I went through my first panoramic life review, I became the people I had hurt physically and emotionally, feeling every nuance of the damage I had caused. However, during my next two reviews, I was able to experience the peace and enormous gratitude of my hospice patients when they took their last breath in my arms. You know the story of the Book of Judgment? Guess what. When you have your panoramic life review, *you* are the one doing the judging. And believe me, you are the toughest judge you will ever have.

It is also important to understand that this judgment has no punishment attached to it. Once you experience the reliving of your life and judge what you think you could have done differently, it's finished. You immediately discard the memory of all guilt, sadness, and regret. You see, in the final analysis, very few things from the life left behind remain important. What matters are not the mistakes you made; what matters to Spirit is how often you were willing to help others through your love, kindness, and compassion. The love you gave away and the potential for love that you instilled in others are the uniqueness of your divine mark. It is only love that you will bequeath the world with your passing, and it is only love that you bring Home with you as your contribution to the expansion of divine consciousness. Only love is real, both here and in the Hereafter. Making a difference in the lives of others is the spiritual foundation of our human existence.

To inspire others to love, to encourage them to dream, and to empower them to keep hope alive are among the most blessed of all achievements. Our simple, spontaneous acts of kindness make the greatest impression on Spirit. A smile given to a stranger, a pat on the back for a discouraged friend, or a meal prepared for someone who is ailing—these are the true marks of compassion in action. Learning to live and love from this place of innocent virtue is a goal we must all set for ourselves. As we align our hearts and minds with the energy of selfless giving, we succeed at attaining the ultimate goal, for in so doing, we assist in raising the level of spiritual consciousness to take us into higher levels of Heaven. The greater the number of spontaneous loving moments we perpetuate here, the higher the levels of consciousness we will inhabit when we reach the Hereafter. The panoramic life review's sole purpose is to act as an impartial tool to help us measure the spiritual growth that took place in the life we are leaving. It determines which threshold of consciousness we are prepared to cross over as we reenter Heaven.

What I have found so fascinating since my own panoramic reviews are the many ways they can enrich our lives on this side of the veil. By consistently implementing the same system of balanced analysis used in the heavenly life review, we can manifest a state of living empowerment. With full knowledge of how the system works, we have the power to refine our destiny day by day. Having this cosmic calibrator at our disposal is a priceless gift, but only if we have the wisdom to put it into practical application. I advise you to consciously seek opportunities to extend your love and share your laughter. Envision yourself as guardian or steward of the Earth, working to protect and nurture global environment. Adopt a homeless animal. Commit a portion of your time weekly to volunteer work. These are just a few of the myriad simple loving actions we can consciously perform to guarantee that we have the most positive effect on all levels of consciousness surrounding us. Because your higher consciousness has guided you to read this information about the panoramic life review, you have been given a great gift.

You also have a blessed responsibility, for much is expected of those to whom much has been given. Employ the panoramic life review method in your daily life, consistently and judiciously, for the greater good of all. Remember, on the other side, you will have the opportunity to become every person you have ever loved or harmed. Knowing that, how will you change the way you treat the people and animals or even the plants and possessions in your life? All of us need to give some serious consideration to this. No thought, word, or deed goes unrecorded in the universe. Therefore, it behooves us to act from a loving heart and an open mind before we have to watch the DVD of our life story.

There is one more incentive for working on a daily basis with the life review system. It endows us with yet another way of circumventing the blue-gray place of spiritual stagnation. Over the years, I have become fixated with the reality of this harsh terrain between the worlds. Clearly, it frightened the heck out of me to realize I was constructing my own superhighway to take me straight to Blue-Gray Town. Each mile of that highway was made of my apathy, bitterness, and cynicism. Now that I am conscious of what I was doing to myself, I want to do all I can to ensure that you do not make the same mistakes. The blue-gray place is a futile state of sustained limbo that is completely avoidable. Through the practice of a daily life review, we can operate in the astral planes.

Becoming conscious of our thoughts, words, and deeds while paying attention to the influence they have on life energy all around us is to step into our divinity one day at a time. In this simple way, we help to create the link between Heaven and Earth. This is the major reason I do hospice work. It is a divine act of love in human form. I give of my strength and courage, as if I were a divine being. At the bedside, I act to protect and comfort the dying. I make sure they are not alone. If they are able, I lead them through a life review in order to open their heart and lose the fear of the impending transition. I make a difference at the bedside, not only for the person leaving this world, but also for the family left behind. My panoramic life review showed me that this is my unique divine mark. Incorporating the life review into your daily schedule of spiritual practices will offer you the same insight, and so much more.

Now let's take a good look at how to perform the daily life review. I find the moments right before bedtime to be the best time to practice life review exercises. I make sure that I am ready for bed, since the review often lulls me into a deep sleep. In an effort to set the right tone for the review, I play New Age music with a heavenly rhythm. Soft candlelight and aromatherapy in the scent of lavender add to the atmosphere. Of course, these celestial aesthetics to create heavenly ambiance are completely a matter of personal preference (although if you are as ADHD as my wife says I am, at least try a couple to help you concentrate). Next, I usually do a few light stretches. It is important to clear the mind of clutter just as you would before meditation, so take a few deep conscious breaths. Now you are ready to begin the nightly life review.

- Think through the day and review the spontaneous events that happened for which you could be considered a blessing. (For example, did you open the door for someone, or put extra money in the parking meter for the next car?)
- Think over the spontaneous events of the day when someone else was a blessing in your life. Give thanks and take a very deep breath. You have now activated your spiritual recognition of the flow of divinity from the higher dimensions into this one.
- Assess the events you feel you could have handled differently. Put yourself in the other person's place to understand the other point of view. If applicable, forgive yourself and surround the situation with light. Assert to make any necessary amends.
- Say your prayers (your willful, conscious intentions) and go to sleep.
- First thing in the morning, count your blessings. A conscious state of gratitude creates a vibratory magnetic field around you that will serve to attract more blessings throughout the day.

I have also found that making to-do lists helps to keep me on track. Of course, I got the idea from all the lists my wife writes for me but, nonetheless, it works. Our minds need to be trained to look for opportunities to help others. To that end, making a daily list of the things we wish to accomplish can be a tremendous asset. Our spiritual life is as important as anything we do to physically maintain ourselves and the world we live in. Taking the time to practice the daily life review will not only deepen your connection to the upper realms, but it will also produce a flow of joy and beauty in your life unlike anything you could imagine. The immutable universal laws of attraction and correspondence are active every minute of every day. We are constantly attracting people and circumstances into our personal reality based on the thought forms we project and the loving energy we extend. The same holds true for unloving energy. These energies accumulate over the duration of our lives so, during our transition, the panoramic life review brings us face-to-face with the actual effects they produced. We can begin today to secure the fact that, when the moment of truth arrives, the life we review will be one filled with dazzling snippets of service to the spiritual evolution of All That Is.

CONCLUSION

If It's True, What Shall We Do?

The more we give love, the greater our capacity to do so.

—Dr. David Hawkins

My willful, conscious intent in writing this book was to share the knowledge and insight I have gained throughout my life and as the result of my three deaths. My motivation in sharing this knowledge is your empowerment. I want you to know that empowerment is within your grasp. The pillars of the Fourfold Path to Power (love, choice, belief, and prayer) work best to improve the quality of your life, both here and in the Hereafter, when used daily. In addition, I believe that the conscious understanding and application of The Seven Lessons will instill within you a sense of meaningful potential that will inspire you to live your life with genuine vitality, purpose, and gratitude.

A main goal in our lives must be to increase our ability and capacity to love. From all I have learned on the other side, I honestly feel this is most easily accomplished through applying (either by teaching or by learning) the pillars and lessons in every situation we find ourselves in. Because I now comprehend life as a university designed for higher learning, I also understand that obstacles to our ability to love will probably be thrown at us. But through this, we inevitably grow stronger, both emotionally and spiritually.

It is imperative that we do not fear the challenges we have attracted to promote our growth. Armed with gratitude and good cheer, there is no mountain we cannot climb in order to reach the pinnacle of our spiritual power. Every obstacle we encounter is an opportunity for spiritual evolution.

Once we have attained this level of higher awareness, the next step in our spiritual edification is to enhance our gifts from Spirit—such as psychic and intuitive abilities. It is necessary to our success to make a list of the gifts we wish to obtain. And let me tell you from experience, we *can* obtain these types of gifts. Do you want to be telepathic? Do you long to commune once more with your departed loved ones? Do you wish to attune yourself to the higher dimensional frequency of the Crystal Cities? Or do you want to communicate with animals?

No matter what your personal aspiration might be, it is important to articulate it and to have it firmly planted in your heart and mind. Determined mental and emotional focus will invariably manifest the achievement of your desire more quickly than anything else. Understand that your thoughts are mighty forms of energy. Focused thoughts are even more powerful and can manifest with greater velocity. This divine method is at your disposal for the sole purpose of fulfilling your dreams and desires. On a daily basis, get into the habit of focusing your positive energy on what you wish to manifest instead of putting energy into thinking about anything you do not wish to encounter. Write your desires down on a piece of paper or a note card and place it where you will see it each day.

Please remember this: that which you seek, seeks you. Every burning desire within your soul is a sacred aspect of your divine mission. It is a gift that Spirit planned for you to cultivate and reveal, for the greater good of all. Nothing we do in life is done merely for self-gratification. From the highest spiritual perspective, all actions in life are executed for the advancement of the collective consciousness of life, whether you are aware of this or not. Furthermore, the collective consciousness I am referring to is universal and therefore includes all other dimensions and realms. This understanding lends itself to the realization that there is no real time/space differential. All time is happening in the Now. Through this insight of quantum mechanics, it is possible for us to communicate with the other side, for in reality, we have never separated or disconnected from Heaven. An aspect of our eternal self is still there, and love acts as the universal conductor through which we can make contact with the other side.

As a result of my many years in hospice work, I have found a highly effective system of transdimensional communication. Fortunately for all of us, it is also very simple. You see, our success in making contact with departed loved ones is predicated upon our motivation for wanting to do so. If we are inspired by selfishness, greed, anger, or guilt, chances are we will fail in our attempts to reach those in the Hereafter. Negativity holds no place or power in the higher realms. In order to successfully lift the veil between the worlds, a spirit of love and celebration needs to lead the way. A spirit of joy and lightness will bridge the gap between here and the afterlife far more quickly than a spirit of concern or sorrow.

Concentrating on the love and laughter we shared with the one we wish to contact is the key to our success. The higher realms and surrounding dimensions can feel the vibratory quality of the energies we project. Love and light are able to penetrate all levels and dimensions. Grief, guilt, and anger are heavy energies that only tend to thicken as they leave the physical realm. Thus, they fall short of the mark, never reaching the higher realms.

You can empower yourself to reach a loved one beyond the veil. First, be patient with this process, for it takes time and practice to perfect. In many ways, this practice is similar to those steps followed in the panoramic life review; the major difference is a conscious change in your intention. I suggest sitting quietly either at sunrise or sunset because the veil between the worlds is the easiest to penetrate at dawn and dusk.

Place a crystal or clear glass bowl of distilled water on a white (representing Spirit) or purple (representing the higher realms) cloth in front of you. The water acts as an electrical conductor to help your loved one's spirit navigate between the dimensions. Light a birthday candle (representing the vital energy of human personality) in the favorite color of the person you wish to contact. If you are not sure which color that is, choose your own favorite color or white—the color of purity. You may also play soft music and burn incense to augment the nature of this spiritual exercise. Place a picture of your departed loved one before you or hold a favorite possession of his or hers in your hand, and pay particular attention to what this person brought into your life. The act of focusing on those things for which you are most grateful deeply enhances the process of transdimensional communication. Concentrate upon the flame to help you hold your focus. Take a few deep, cleansing breaths, pay attention to the rhythm of your breathing, and simply relax.

Next, begin to relive some of the most treasured memories you have shared with this person. Remember the special occasions filled with laughter or the family holidays brimming with love. These are moments forever etched in time, and they possess a vibratory imprint capable of reaching across the distance to beckon your loved one to you. As you hold these mental and emotional pictures firmly in place, the communication begins. Initially, you may only experience a glimpse of his or her face or perhaps a flash of a shared moment you had forgotten. When this occurs, you know you have made contact. From that point on, with consistent practice, a glimpse becomes a moment, and from a moment, eternal communication is established.

Do not be surprised if you experience physical sensations, such as a chill in the room; the extinguishing of the candle; or the scent of cologne, perfume, or flowers. You may hear chimes from a faraway place. And in many instances, after diligent practice, the loved one may actually appear before you.

Over 66 percent of all people who have lost a loved one experience some form of communication. These steps, if followed exactly as I have outlined, will enhance your ability to communicate with the other side.

For those of you who would rather see the future than talk to the dead, or would like to read the Akashic Records in hopes of accessing the history of the world, there are a few different steps to follow. First, analyze your reasons for seeking this gift to make sure they are not just self-serving. You must be willing and prepared to use the knowledge you gain for the improvement of our collective consciousness.

Then imagine a friend or a loved one who you know is in trouble (under great emotional stress or physically ill, for example). Ask your higher self to show you what lesson that person is trying to learn from the problem and what you need to know to help her (or him). For me, the answer comes along like a preview at the movies—I see a quick video of the person's life, showing me what she is going through, along with the cause of her problem. However, there are many unique ways to intuit, and there is no wrong way. You may hear a voice or sense the answer without pictures. The point is to trust in the process, to trust your intuition in whatever form it shows itself. The more you follow your intuition from a place of purity, for the benefit of others, the stronger it grows. It is mandatory for you to understand that a strict code of honor and integrity must be followed for those who venture into the realm of spiritual powers and gifts.

In the absence of such spiritual honor and integrity, it is easy to succumb to the temptation of effortless personal gain. Many people get trapped by avarice and self-interest. When coming from this place of egocentricity, the intuitive gifts, though innate, may be blocked. Often you may find yourself being pulled into the lower astral planes as the result of trying to exploit your intuitive gifts for less than the highest spiritual ideal. Believe me, in the early days right after the lightning, I learned this lesson the hard way through many misguided adventures. I also believe one of the reasons for the events of the third near-death experience, my trip to the blue-gray place, was to show me the consequences of my misuse of the enormous spiritual power and knowledge I had been given. I implore you not to make the same mistakes. The gifts of the Spirit need to be used in ways that increase the ability and capacity for manifesting more love in the world.

In closing, I would like you to visualize yourself as a mighty, powerful spiritual being, surrounded by many other spiritual beings, standing on a bluff above a sparkling, clear, calm lake. You hold a handful of precious jewels that represent your soul's highest dreams and wishes. The waters signify the collective consciousness of all life. Imagine yourself casting the jewels upon the lake and visualize the ripples casting across the water. Then see the spiritual beings around you, one after another, casting their jewels into the water until, for as far as you can see, the entire surface of the lake is intricately etched with the crossing paths of your combined spiritual ripples.

This is the way life operates. Your every thought, word, and action has an effect on, and is affected by, the thoughts, words, and actions of everyone else. While holding this knowledge close to your heart, please make a conscious effort each day to create your ripples from jewels of love and for the express purpose of contributing to the evolution and empowerment of our entire spiritual family. Unity consciousness is the ultimate goal of this era in human history, and coming to terms with the reality of our spiritual power is the surest way of reaching this goal. It doesn't take a trip to Heaven for you to help create peace on Earth. It only takes the conscious efforts of a compassionate heart to make the world a better place—one loving act at a time. Spirit expands as we do. Every act of kindness, forgiveness, and compassion helps the universe create more love, for only love is real, and as such, love is the vital force that gives birth to, nurtures, and sustains all life. Love is our essence as well as our destiny. When we elevate our consciousness to embrace this truth, when we align our every thought, word, and deed with its vibration, then we shall create Heaven on Earth.

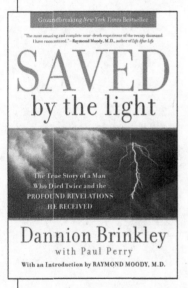